Getting Started as a Consulting Engineer

getting*started*

as a

CONSULTING ENGINEER

Professional Publications, Inc.
Belmont, CA 94002

In the ENGINEERING REFERENCE MANUAL SERIES

Engineer-In-Training Reference Manual
Engineering Fundamentals Quick Reference Cards
Engineer-In-Training Sample Examinations
Mini-Exams for the E-I-T Exam
1001 Solved Engineering Fundamentals Problems
E-I-T Review: A Study Guide
Diagnostic F.E. Exam for the Macintosh
Fundamentals of Engineering Video Series:
    Thermodynamics
Civil Engineering Reference Manual
Civil Engineering Quick Reference Cards
Civil Engineering Sample Examination
Civil Engineering Review Course on Cassettes
101 Solved Civil Engineering Problems
Seismic Design of Building Structures
Seismic Design Fast
Timber Design for the Civil P.E. Exam
Fundamentals of Reinforced Masonry Design
246 Solved Structural Engineering Problems
345 Solved Seismic Design Problems

Mechanical Engineering Reference Manual
Mechanical Engineering Quick Reference Cards
Mechanical Engineering Sample Examination
101 Solved Mechanical Engineering Problems
Mechanical Engineering Review Course on Cassettes
Consolidated Gas Dynamics Tables
Fire and Explosion Protection Systems
Electrical Engineering Reference Manual
Electrical Engineering Quick Reference Cards
Electrical Engineering Sample Examination
Chemical Engineering Reference Manual
Chemical Engineering Quick Reference Cards
Chemical Engineering Practice Exam Set
Land Surveyor Reference Manual
1001 Solved Surveying Fundamentals Problems
Land Surveyor-In-Training Sample Examination
Engineering Economic Analysis
Engineering Law, Design Liability, and
    Professional Ethics
Engineering Unit Conversions

In the ENGINEERING CAREER ADVANCEMENT SERIES

How to Become a Professional Engineer
The Expert Witness Handbook— Guide for Engineers
Getting Started as a Consulting Engineer
Intellectual Property Protection— A Guide for Engineers
E-I-T/P.E. Course Coordinator's Handbook
Becoming a Professional Engineer
Engineering Your Start-Up
High-Technology Degree Alternatives
Metric in Minutes
Engineering Your Job Search

GETTING STARTED AS A CONSULTING ENGINEER

Printed in the United States of America

ISBN: 0-932276-53-9

Professional Publications, Inc.
1250 Fifth Avenue, Belmont, CA 94002
(415) 593-9119

Current printing of this edition: 6

Revised and reprinted in 1995.

PREFACE

At one time or another in their careers, most engineers dream about becoming consulting engineers. After gaining experience in their particular fields, a fraction of these engineers recognize that their abilities and knowledge would be useful and valuable to a variety of customers. A smaller fraction of these engineers take the big step and set out on their own.

As an engineer, you are lucky to work in an environment that is ideal for independent consultants. There are only a few professional areas in which consultants satisfy a significant part of the demand for services in those areas. Independent engineers, however, are employed in the normal course of many engineering projects.

There is another factor that makes engineering an ideal profession for consultants. There is significant demand for engineering services in all areas. Engineering consultants design processes and equipment, perform tests, and analyze accidents. Independent services are needed from all engineering disciplines, including mechanical, electrical, civil, chemical, and industrial engineering.

Getting Started as a Consulting Engineer has been written for those engineers who can and want to do more than just dream about working for themselves. The decision to become independent should be made on the basis of expected changes in income, prestige, independence, and personal satisfaction. This book discusses these issues, and asks you questions that you need to answer before deciding to become a consulting engineer.

Not every engineer can be or should be a consultant. However, that is not a valid reason for you to dismiss the concept for yourself. **Getting Started as a Consulting Engineer** will help you turn your dream into reality.

TABLE OF CONTENTS

SHOULD YOU BECOME A CONSULTING ENGINEER?

The constant aim of the consulting engineer, like that of any other independent businessman or professional in private practice, is success. The goal of this book is to give you information, guidelines, and pointers which will help you to achieve the success you are aiming for.

The first step in deciding whether to become a consulting engineer is to assess your own situation. Is an independent consulting practice right for you? If it is, are you ready to get started? The prerequisites for setting up an independent engineering consulting practice fall into three major categories: professional, personal, and financial. Let us look at them one by one.

PROFESSIONAL QUALIFICATIONS If you wish to establish a consulting engineering practice, you must, before all else, become registered as a professional engineer. Without registration and the accompanying license to practice, you may not legally take responsible charge for engineering work. If you are not already registered, the first step towards independent consulting is to become registered.[1]

Legally speaking, you may set up a consulting firm as soon as you obtain your license to practice as a professional engineer (P.E.). Most prospective clients however, will look for a record of experience and accomplishment before engaging your services. If you have become registered as a P.E. after only the minimum amount of experience, it may be to your advantage to work in your field for a few years before

---

[1] *How to Become a Professional Engineer*, available from Professional Publications, Inc., is a thorough, up-to-date guide through the registration process.

starting your consulting practice. In this case, you should try to take on as much responsibility in as many phases of engineering work as possible within your job, so that your experience will be maximally useful to you later. Meanwhile, you can lay some of the groundwork for your future consulting practice.

If you are ready, professionally speaking, to become a consulting engineer, you should take a hard look at both your personal and financial readiness to set up your practice. In doing this, the most important thing to remember is that your new consulting engineering practice will fundamentally be a small business, even though you will be offering professional services. You will not be selling a tangible commodity, but you will be selling your time and talent. Your practice will make the same personal and financial demands on you that any other small business would make. Most of the advantages and disadvantages that apply to small businesses in general will apply to your practice as well.

PERSONAL ADVANTAGES OF OWNING YOUR OWN BUSINESS

Having your own consulting business can bring you a number of benefits that are not available to those who are employed by someone else.

Profits. In the long run, if you are successful, you will probably enjoy a higher income than that which you could hope for as an engineer employed by a firm. As an employee, your salary depends not only on how well you do your job, but also on company policy, seniority, and other factors not connected with your performance. As an independent consultant, your income is directly dependent on your performance, with no preset upper limit.

Independence. One of the most universal motivations among small business owners, including consultants, is the desire to be one's own boss, to be free to decide what to do and how to do it. As a consulting engineer, you will have the authority to make all decisions about your practice.

Creative Freedom. As the person in charge, you will have the freedom to express your creativity, both in your engineering work and in the solutions you devise to meet business problems. You will not have to "go through channels" or defer to company traditions.

Pride in Accomplishment. You can take pride in knowing that the success you enjoy in your consulting business has been accomplished through your own talent and efforts.

DISADVANTAGES OF OWNING YOUR OWN BUSINESS Although the advantages of owning your own business are great, the disadvantages are just as real. In deciding whether to go into independent consulting, you will need to weigh the advantages and disadvantages according to your own situation.

Risk. If your business succeeds, you will prosper. However, if it fails, you could lose your entire investment. Not only that, if you are sole proprietor or a partner, you could lose your personal assets as well.

Uncertain Income. You will have no guaranteed monthly income. Rather, you will be dependent on the ups and downs of the business and its seasonal and other fluctuations. This is particularly true during the early years of the business.

Responsibility. While having the ultimate responsibility for your business can give you the freedom to make decisions and express your creativity, it can also have its negative side. If you make a wrong decision, there is no one else to blame, and you must take responsibility for employee errors as well.

Effort. Because of the unrelenting demands of the business, you will probably have to work long hours (twelve hours a day, six days a week is not unusual), especially during the first few years. It is almost impossible to leave your work at the office; there is always something more that needs to be done.

Investment. To get your practice started, you will have to make a considerable investment from your personal assets, and you may have to borrow money as well. It may be a number of years before you recover your initial investment.

Lack of Flexibility. It may be virtually impossible to take any time off, even for a few days' vacation, for several years. Also, having made the investment, you cannot easily change your mind or your job. If you find that you do not enjoy consulting work, it may take considerable time and involve considerable loss to get out of your practice.

THE PERSON-ALITY OF THE INDEPENDENT CONSULTANT   Before deciding to start your own consulting business, you should consider the match between the requirements of the business and your own personality characteristics. Try to answer each of the following questions honestly. These questions have no right or wrong answers. They are simply intended to help you decide whether you are likely to enjoy and be successful in a consulting career.

Are you a self-starter? Do you set your own goals, establish your own deadlines, and work toward them without being pushed or guided by others? Or do you work better if someone else structures the situation for you? Or do you prefer to work as little as possible, doing only the minimum to satisfy others' demands?

As an independent consultant, you must be a self-starter, especially if you work alone. There will be no one but you to set the goals and establish the deadlines. If you prefer more structure, you might consider working with a partner who likes to take the lead. If you usually try to do only the minimum, an independent practice is probably not for you.

Do you like other people? Can you get along with most people? Or do you prefer to associate with the people you already know and like? Or do most other people get on your nerves?

As a consulting engineer, you must deal with a large variety of people, from clients to secretaries, from bank officers to construction workers. Even more than other businessmen, who have tangible products to sell, you must rely on good relations with others for the success of your business. If you are very shy, or if you basically don't like other people very much, it could be an impediment to your success.

Can you lead other people? Can you usually get others to go along with your ideas? Or do you have to push and struggle to get others to do what you want? Or would you rather leave it to someone else to get things moving?

Although you may start out as a one-person operation, you will soon find that some leadership ability is called for. Whether it is enlisting the cooperation of a client's employees or supervising your part-time bookkeeper, many aspects of independent consulting require the ability to get other people to do things willingly. If leadership is a strain for you, or if you prefer to leave the leadership role to others, you may experience difficulties in keeping your business moving.

Can you take responsibility? Do you like to take charge and see things through? Or would you rather let someone else take responsibility, only taking over if necessary? Or do you dislike responsibility and resent the people who do take charge?

Willingness to take responsibility is an important trait for the consulting engineer, as is the ability to get yourself started. If you prefer to share responsibility, you might be able to succeed with the right partner. However, if you look upon others who take charge as "eager beavers" and show-offs, you will have difficulty both in taking responsibility yourself, and in letting a partner take it. The results could be disastrous for your business.

Are you a good organizer? Do you enjoy planning, and do you usually have a clear idea of what you are going to do before you do it? Can you get reorganized quickly if something unexpected comes up? Or do you start out organized and then give up if things don't go according to plan? Or do you just take things as they come, without thinking too far ahead?

There is an old adage in business that says, "Failing to plan is planning to fail." If you are poor at organization and planning, or if you dislike planning, your consulting practice will suffer.

Are you a hard worker? Are you willing to work as long and hard as necessary to get something done? Or do you work hard up to a certain point and then quit? Or do you think that hard work is just a waste of effort?

Getting a consulting business started requires you to work long hours and to do many things (such as advertising and accounting) for which you may not be trained and which you may find very tedious. If you give up when the going gets tough, or if you can't see the point of working hard anyway, independent consulting may not be a wise choice for you.

Are you decisive? Can you make up your mind quickly, and usually make a good decision? Or do you need time to make up your mind, and usually regret decisions that you make in a hurry? Or do you dislike making decisions at all?

When you are the boss, the decisions are yours to make. If you dislike decisions, or if every decision takes a long period of pondering the alternatives, you will probably not be very happy in the role of consulting engineer.

Can people trust what you say? Can you be relied on to mean what you say and stand by it? Or do you sometimes say something you do not mean because it is the easiest thing to say in that situation? Or do you think that it doesn't matter if the other person gets fooled?

In business in general, a reputation for honesty and reliability is essential. However, as consulting engineer, you are both a businessperson and a professional. This means that you have both a business and a professional reputation to maintain, requiring the strictest standards of honesty and ethical conduct.

Are you persistent? If you make up your mind to do something, do you stick with it no matter what? Or do you usually finish what you set out to do, provided there are no serious obstacles? Or do you give up as soon as you encounter difficulties?

Obstacles and difficulties are a normal occurrence in business. If you get discouraged and give up easily, you will not stay in business for very long. Determination and persistence are very important factors in business success.

Are you healthy and energetic? Do you have lots of energy, and never feel run down? Or do you have an average amount of energy? Or do you run out of steam faster than most people you know?

Because of the demand for hard work and long hours, you will need at least an average amount of energy to succeed as a consulting engineer. A high energy level and good general health are an important asset to your business.

If the personality of the successful consultant had to be summed up in one phrase, that phrase might be "resolutely positive." To be successful as a consulting engineer, you must be ready to look on the bright side of almost everything, from human nature to unexpected delays to long hours of unpaid labor. With this positive attitude, you can meet and overcome almost any adversity, but without it, you can easily become discouraged. Before deciding to commit yourself to a consulting career, try to assess yourself honestly to see whether you have this persistently positive attitude toward life.

FINANCIAL CONSIDERATIONS If you are professionally qualified, and if you believe that your personality is suited to the demands of a consulting practice, the next question is: Can you afford to go into business?

Fortunately, since you do not need to acquire an inventory, the actual outlay of cash for getting started as a consulting engineer can be kept to a minimum. However, much more than just startup costs are involved. It is a rare new business that does better than break even in its first year of operation, and most businesses do not show a profit until the end of the second or even third year.

This means that you must have reserves or income from other sources that will be sufficient to meet not only your business expenses but also your personal and family expenses for at least one year, and more likely two or three. Many potentially successful businesses and consulting careers are abandoned every year because of inadequate financial foundations. It is better to wait a few years, building up a secure financial base and improving your chances of success, than to plunge in with only your energy and ideas to back you up, risking an early failure.

PLANNING: THE KEY TO SUCCESS IN BUSINESS

Before setting up your consulting firm, a great deal of planning is in order. If you look at statistics, you will see that the odds appear to be against your success. Planning can turn the odds in your favor.

Every year, thousands of new businesses are started in the United States. According to statistics from the Small Business Administration (SBA), more than half of new business failures fail in their first few years, with most of the failures occurring in the first year. The majority of these failures are attributed to poor management, but there is good reason to believe that poor management is a result of poor planning.

Studies comparing new businesses that survived their first three years with new businesses that failed in the first three years have revealed a number of differences. Most of these differences have to do with planning and expectations. For instance, owners of failed enterprises have been found to spend, on the average, less time in preparation and research than owners of successful businesses. A much larger percentage of successful owners than failed owners have taken business courses and regularly read business-related books and magazines. Many more successful owners than failed owners have sought professional advice from lawyers, accountants, and others before starting up. Most successful owners have been found to expect a slow start and low income in the beginning, while failed owners tend to have expected quick success and a high income within a year or two.

These findings show that success or failure is not just a matter of luck, but that planning can greatly increase your chances of success.

Your planning should focus on competence in three major areas—technical, marketing, and financial. In each of these areas, you will need to assess your current knowledge and capabilities, and then take steps to make up for the deficiencies.

PLANNING FOR TECHNICAL COMPETENCE If you are already a professional engineer, your competence is at a level that your state is willing to recognize with registration and a license. However, you need to decide exactly what engineering services you intend to offer your clients. As a consultant, you may need to perform a much wider range of tasks than you do in your current job.

Are you fully competent in all phases of the work you intend to do? Do you have experience in all phases? If there is any doubt in your mind about being able to satisfy the needs of your prospective clients, you should eliminate those doubts by taking positive action to gain experience in all the phases of engineering work that you expect to perform as a consultant. Also, you should read and study in the areas that you feel you need to strengthen.

During the planning stage, you should talk with established consulting engineers in your field. Not only will a network of professional acquaintances be helpful to you later in your practice, you can benefit from their experience at an earlier stage, in planning the services you will offer.

In addition to engineering competence, you will need competence in communication skills. A sizeable proportion of the consulting engineer's time is taken up with presentations, discussions, and writing letters and proposals. To a large degree, the success of a consulting business depends on the consultant's ability to get ideas and information across to others clearly, succinctly, and persuasively. If you feel any uncertainty about your ability to carry out the communication tasks of a consulting practice, particularly the writing tasks, you may wish to polish your writing skills by taking courses in business English or proposal writing.

PLANNING FOR MARKETING COMPETENCE If you have always thought of engineering as a profession, the word "marketing" may have an undignified ring to your ears. One prominent writer in the field of consulting engineering prefers to say "business development" rather than "marketing," while others speak of "generating referrals." Consulting engineering is both a profession *and* a business, and it is important to recognize that marketing is essential to any business, even one that offers only professional services.

Marketing should not be confused with advertising. In a very real sense, *marketing is the relationship between the consultant and the outside world.* A marketing plan encompasses many different aspects of that relationship, including the following:

- a specific definition of the services you will offer
- a definition of your target population of potential clients
- a plan for reaching that population
- an effective message conveying your services to potential clients
- pricing your services at the optimum level

To maximize your chances of success, you should make definite and specific plans for each of these aspects of marketing. Taken together, these plans will constitute your marketing strategy. Your marketing strategy, in turn, will be one of the main determinants of the success of your consulting practice.

PLANNING FOR FINANCIAL COMPETENCE

Financial competence has several components, including the following:

- an adequate financial base for survival during the start-up phase
- adequate knowledge and use of accounting procedures and records
- knowledge of how to use accounting information
- techniques for minimizing costs
- techniques for minimizing losses
- precautions to reduce the impact of unexpected setbacks

The time to think about each of these financial considerations is now, before you establish your practice, not later, when you begin to run into problems. Many things can only be fully learned through experience, but you can learn from experience much faster when you have some theoretical knowledge to apply to a concrete situation. Thinking things through ahead of time may allow you to recognize the nature of a problem when it comes up, saving you a period of confusion and false starts.

One of the best investments you can make in your future company, in terms of financial planning, is to read some introductory accounting books or, better yet, to take a course in basic accounting. This will give you a solid basis for understanding the important financial aspects of your practice.

Each of these aspects of planning for competence in your practice will be taken up in greater detail later in this book. This section is intended only to give you an overview of some planning considerations and to emphasize that thorough planning is essential in establishing a successful consulting engineering practice.

DECIDING ON THE FORM OF OWNERSHIP

When establishing a new business, you will have three basic options for type of ownership. These are the (1) sole proprietorship, (2) partnership, and (3) corporation. Each of these forms of ownership has advantages and disadvantages. The decision to choose one of them should be made only after determining that the form suits you and your circumstances best.

SOLE PROPRIETORSHIP Sole proprietorship, also called single proprietorship, is ownership of a business by a single person. It is the most common form for small businesses. More than one-third of small consulting engineering firms (those with fewer than 10 employees) are sole proprietorships. Simplicity is one of the major advantages of the sole proprietorship.

Sole proprietorship is the easiest form of ownership to set up. Other than the necessary licenses, which apply to all forms of ownership, there are no legal formalities involved in establishing a sole proprietorship.

You may legally practice engineering in any state as a sole proprietor if you are registered as a professional engineer in that state.

The sole proprietorship allows you to use the simplest type of accounting system, particularly if you have no employees.

In addition to simplicity, the sole proprietorship offers other attractions as well, such as a high degree of independence. A sole proprietorship allows you virtually total control of your business. As sole owner, you will have sole professional responsibility for all the work you undertake. All supervision and management will be your responsibility, and you can make all decisions for the business.

There are also disadvantages to the sole proprietorship form.

Perhaps the greatest disadvantage is your personal legal and financial liability for all acts of the company. If you fail to pay company debts on time, your personal assets (home, car, savings, etc.) may be seized or attached. In the event of a legal judgment against the company, your personal assets are again at risk.

The total independence of the sole proprietor may be attractive, but it may be burdensome as well. In times of crisis or trouble, there is no one to share the responsibility or to help make the decisions.

There is also the problem of continuity. At the beginning, your whole concern will be with the sheer survival of your firm. Later, when it is a successful, ongoing enterprise, you may become concerned for its future. In the event of your protracted illness, absence, retirement, or death, the sole proprietorship form makes no provision for the continuation of the firm as an independent entity.

It is often said that the sole proprietorship is disadvantageous in terms of taxes. This is because income to the business (after expenses) is taxed as personal income from self-employment of the owner. However, tax advantage or disadvantage depends on a great many factors; the form of ownership is only one of them.

While the form of ownership does influence the basis and rate of taxation, it must be considered in conjunction with all the other influencing factors. Circumstances such as the volume of business you anticipate, the number of people you employ, and the amount of personal assets that you wish to protect from business liability will affect the basis and rate of taxation. It is best, therefore, to seek advice from a tax accountant before deciding on the form of ownership.

PARTNERSHIP   Legally, a partnership (ownership by two or more persons) is very similar to sole proprietorship, and is also a common form of ownership for small businesses. About 12 percent of small consulting engineering firms (those with fewer than 10 employees) are partnerships.

Partnerships have many of the same advantages and disadvantages as sole proprietorships. A partnership firm can be set up with the same minimum of formalities as a sole proprietorship. The practice of engineering by partnership firms is legal in all states. Taxation of partnerships is on essentially the same basis as taxation of single proprietorships. Each partner has personal responsibility and liability for the acts of the firm, just as sole proprietors do.

The additional owners in a partnership bring some additional considerations with them however. Although establishing a partnership requires no legal formalities, it is best to draw up a partnership agreement before opening the business. Neglecting to do so can lead to misunderstandings, disagreements, and even lawsuits later on. If you are considering setting up a consulting engineering firm with one or more partners, you should obtain the services of an attorney in preparing a written agreement. For maximum protection, each partner may retain a different attorney.

The partnership agreement should state clearly the amount and kind of contribution to be made by each partner, including the percentage of the firm that will be considered to be owned by each partner. It should specify the basis on which profits and losses will be divided among the partners, and when and how profits may be withdrawn. It should state the basis for compensation of the partners (services, investment of capital, etc.) and how compensation will be made. The document should make clear provisions for changes in ownership, such as adding new partners, the withdrawal of a partner, changes in the amounts of partners' investments, and the like.

Although some of these issues may seem distant or irrelevant in the beginning of a partnership, thinking them through and arriving at an agreement ahead of time is extremely important. Only with a written agreement can you ensure that each partner's interests are adequately protected and prevent the pain and disappointment that can result from misunderstandings.

Under some circumstances, a limited partnership may be the most desirable form of ownership. A limited partnership differs from a general partnership in that one or more of the partners is "silent"; that is, a limited partner makes a financial contribution to the business, but plays no role in its management. While general partners have unlimited personal liability, limited partners are generally liable only to the extent of their investment. Although partnership agreements are desirable in general partnerships, they are legally mandatory in limited partnerships. In addition, a limited partnership firm must file a certificate and fee with the state.

Perhaps the most obvious advantage of establishing a partnership is the increase in possibilities for financing. Pooling the resources of two or more partners can provide a new firm with a more solid financial base than either of them could provide alone. Partners may also find it easier to get business credit.

The concentration of decision-making and responsibility in one person is both a drawback and an advantage for the sole proprietorship. Similarly, the fact that decision-making and responsibility are shared in a partnership has both advantages and disadvantages. Difficult decisions can be discussed and made jointly, and the burden of responsibility may seem less debilitating when there is someone to share it. On the other hand, sharing responsibility may limit each partner's ability to act independently (although each partner is legally an agent who has the right to make binding agreements on behalf of the firm).

Along with responsibility, financial liability is shared in a partnership. This may be helpful, since financial losses shared by two or more people may be less ruinous than losses that must be absorbed by one person. On the other hand, since each partner can enter binding agreements on behalf of the entire firm, the acts of one partner may result in liabilities for the other partner or partners. Also, since a partner's share of the firm is not legally distinct from his personal assets, any partner's failure to pay personal debts may result in the attachment of his interest in the firm. The consequent loss of capital could create serious problems for the firm.

Continuity is a problem of direct ownership which is not solved by a partnership. A partnership is generally considered to be dissolved upon the withdrawal or death of one of the partners. Of course, the firm may be reconstituted by the remaining partners, with or without replacement of the lost partner.

Partnership firms that are successful enough to grow into large enterprises may face the problem of whether or not to make partners of key employees. Accepting new partners may be necessary to avoid loss of talented employees, but it may also increase the number of partners to the point that decision-making becomes unwieldy. As this consideration indicates, there may be very long-term consequences of the choice of ownership form that you make at the beginning.

CORPORATION Rather than being a direct owner of your consulting business, either as a sole proprietor or as a partner, you may prefer to incorporate, or create the firm as a corporation. This means that you will set up a separate legal entity or "legal person" for the conduct of business, including entering contracts, paying debts, and paying taxes. The corporation is the most common type of ownership form for large businesses. It is also the most common form for consulting engineering firms; more than 50 percent of small consulting engineering firms, and more than 80 percent of larger ones, are corporations.

The corporation is a fundamentally different kind of business than a sole proprietorship or partnership, and has a very different set of advantages and disadvantages. In fact, the corporate form is specifically designed to avoid some of the disadvantages of direct ownership.

Because the corporation is a legal "person," its acts and liabilities have some degree of independence from the acts and liabilities of the owner or owners. Income to a corporation is not income to the owners, unless paid as salary or dividends. Debts of the corporation are paid from the assets of the corporation, not from assets belonging directly to the owners. Taxes are paid on income to the corporation, at corporate tax rates. Owners of corporations pay taxes on their personal income that are separate from and in addition to taxes paid by the corporation.

These features of the corporate form of ownership protect the owners from personal liability for debts of the corporation. Even if the corporation should go bankrupt, the personal assets of the owners are not ordinarily liable to seizure or attachment. If you have a large personal worth that you would like to protect against business liability, the corporate form may be your best choice.

The corporation is widely thought to offer tax advantages over direct ownership forms. In general, this is true when profit levels are high. However, as was indicated in the section on sole proprietorship, many factors enter into the calculation of relative tax advantage. The advice of a tax accountant can help you decide which form of ownership will suit your situation best.

The corporation, with its independent status as a legal entity, provides the best guarantee of continuity for your business in the case of your death, incapacity, or retirement.

If you choose the corporate form, you will almost certainly prefer the closely held, or close, corporation rather than the public corporation. That is, all stock will be held by you and a small group of co-investors. "Going public," or offering shares of stock for sale on the stock market, requires legal and financial resources that are usually beyond the means of beginning consultants. Almost all small incorporated businesses are of the closely held type.

In a corporation, there is no necessary connection between ownership and management functions. This means that decision-making power can be vested in executive officers and a board of directors, which can govern by majority vote. This form of management offers distinct advantages over the cumbersome consensus procedures that an extensive partnership may entail.

However, for the principal owner of a close corporation, the distinction between ownership and management may be largely irrelevant until retirement becomes an issue. The choice to incorporate probably will make little difference in your day-to-day activities. If you are the principal owner, you will probably act much like a sole proprietor; if you have one or more major co-owners, you will probably treat the business much like a partnership. In the latter case, written agreements regarding each co-owner's rights and obligations in the running of the business will be just as necessary as in a partnership. The main difference will be in the increased paperwork and accounting load.

Incorporation is more complicated and expensive than setting up a sole proprietorship or partnership. Forms must be filed, fees paid, and notices posted. Because they are legally independent entities, corporations are regulated more closely than directly owned businesses.

For a consulting engineer, a major disadvantage of the corporation form is that several states prohibit the practice of engineering by ordinary corporations. This prohibition is predicated on the need to maintain personal professional responsibility for engineering work in order to protect public welfare. Particularly in large corporations, the division between ownership and management may make it difficult to guarantee the maintenance of personal professional responsibility. If you live in or plan to do extensive work in a state with such a prohibition, the ordinary corporation obviously would not be a suitable form of ownership for your consulting practice.

An alternative to the ordinary corporation is the professional corporation or professional association. Control of this type of corporation is limited to qualified professionals. Shareholders are protected from personal liability for debts of the corporation, as in ordinary corporations. However, in a professional corporation or association, any individual associate may be held personally liable for professional malpractice. States that prohibit the practice of engineering by ordinary corporations generally allow professional corporations to practice, although incorporation in the state may be required.

Another alternative to the ordinary corporation is the Subchapter S corporation, or more simply, S corporation. The S corporation is in most respects like an ordinary corporation. It differs mainly in that it pays out its income to shareholders very much as an ordinary partnership does to the partners, and the income is taxed as personal income, as in a partnership. Shareholders are protected from personal liability as in an ordinary corporation. Because of its particular set of advantages and disadvantages, the S corporation is often preferred by small business owners for the first few years of operation. There are numerous eligibility requirements for filing election of the S corporation option and numerous regulations for subsequently administering it. Under some circumstances, the S corporation can provide distinct tax advantages, but the legal and accounting advice typically required may negate these monetary advantages.

SETTING UP
THE BUSINESS

Suppose, after carefully evaluating your situation, that you have decided that consulting engineering is the right choice for you, and you have decided on the form of ownership that would suit you best. Now it is time to take concrete steps toward setting up your practice.

DO YOU NEED PROFESSIONAL COUNSEL? Before launching your business, you will need to make some decisions that can be broadly classified as legal and accounting decisions. You may wish to get professional advice in making these decisions.

Legal Advice. Strictly speaking, you do not need an attorney's assistance in setting up any of the forms of ownership. Sole proprietorships and partnerships can be established without legal formalities. Even a corporation can be established without legal advice, if you are willing to go through all the procedures yourself. There are a number of self-help guides on the market for entrepreneurs who wish to incorporate as economically as possible.

This is not to say, however, that an attorney's advice would not be helpful. As we saw earlier, an attorney's assistance can be an important part of effective planning. In drawing up a partnership agreement, for example, legal advice is highly desirable. An attorney can also give sound advice on whether or when you should incorporate. Other areas in which legal advice can help to safeguard your interests include consulting agreements or contracts, service contracts, your duties as an employer, liability and insurance, and patents.

Although it is certainly a good idea to consult an attorney when doubtful about these or similar questions, it is important to remember that business decisions are YOURS to make. Your attorney can give you only information and advice; it is up to you to use that information and advice to your best advantage.

Accounting Advice. Similar considerations apply when you seek accounting advice. Unless your own accounting skills are very good and you have the time to use them, you will almost certainly need bookkeeping and accounting services and advice from time to time.

It will help to get off to a good start if your accounting system is set up before you begin full-time consulting. However, your preferences and interests and those of your accountant may be at odds. If you are setting up a one-person consulting business, you will do best with a very simple accounting system. Most accountants, on the other hand, prefer the more complicated systems that they have been trained to manage. You should choose the system that will be most useful for your particular circumstances.

Whether you use a system set up by an accountant or one of the commercially available systems, you should do the day-to-day bookkeeping yourself, or have it done, at least once a week. This will ensure that you have up-to-date information about your financial status continuously available. Having your accounting done on a monthly or quarterly basis may prevent you from seeing important balance-sheet information in time to take corrective action.

An accountant can also provide valuable counsel at tax time, and may save you a great deal of time by figuring your taxes for you. He may also help you to interpret the meaning of patterns in your balance sheet. Once again, however, the accountant can give you only information (in addition to basic accounting services); you must make the business decisions yourself.

FINANCING YOUR BUSINESS   One of the questions that is probably uppermost in your mind is that of financing. How much money do you need? Where can you get it? There are, unfortunately, no easy answers.

Calculating Your Investment Need. The amount you will need depends on a number of factors, such as the minimum personal income that you need, the type of office that you intend to maintain, and whether or not you have partners or co-investors. (Of course, if you are planning to acquire ownership of an existing practice, there are other considerations as well, such as the purchase price and anticipated taxes. In this discussion, however, it is assumed that you are planning to start a new practice, either alone or with a small number of partners or co-investors.) You can estimate the minimum amount that you will need to invest by adding together the personal and business expenses that you anticipate for the first year and subtracting the amount that you can expect from other sources, including other investors.

Even if you are a careful comparison shopper, equipment and supplies will probably cost more than you budgeted for. Until you learn to estimate your time and costs more accurately, the first few jobs you do will probably take more of your time and materials than you planned for. Insurance bills will add up to more than you imagined, and telephone bills may surprise you. In addition, there will probably be a series of minor disasters. For instance, you may discover that your carefully designed brochure has been printed without your telephone number and will have to be printed all over again. Or, your new typewriter may prove to be a lemon, spending more time in the repair shop than on your desk.

It is safe to assume you will receive no income at all from your consulting practice during the first year, and no personal income from it in the second year. This may sound overly pessimistic. Indeed, it is very possible that, given careful planning, you will break even in the first year and begin to realize a profit in the second year. It is best, however, to proceed from a pessimistic—or better, realistic—standpoint because it will give you a small margin for errors and setbacks.

The amount you must invest is determined largely by anticipated costs. Whatever you can do to minimize your costs (including the cost of your mistakes) will reduce the necessary investment. On the personal side, accepting a modest standard of living until your practice takes hold will reduce the amount of money that you must have available. On the business side, you can hold down costs by avoiding most major investments until you have begun to generate enough income to pay for them.

What investments must you make in order to establish a consulting engineering practice? Compared with a business that sells tangible commodities, a consulting practice requires little in the way of business investment. To start with, you will need a place to work, basic office and design equipment (desk, typewriter, telephone, drafting table, filing cabinet, etc.), business cards, brochures, and some stationery. Until your practice is producing enough income to justify extras, these items essentially are all you will need.

Economizing on Initial Investments. It is both possible and desirable to economize on each of these items. For instance, instead of leasing or renting a suite at a fashionable business address, you should consider renting a single office. Rather than committing yourself to a long-term lease, try to rent your office space on a month-to-month basis. In many cities, it is possible to rent desk space alone, which may include telephone answering and limited secretarial services. This arrangement will reduce the amount of equipment that you will need to buy.

Most economical of all is to work out of your own home. If you have a room that you can convert into an office, this may be an ideal solution for the first few years. For a consulting engineer, office location is not a determining factor in success, so it is best to choose the arrangement that is both economical and compatible with your needs.

It is possible that you will try more than one situation in the first few years, which means that your address may change. If you are concerned about keeping a stable address, you may use a post office box number. Many consultants use a post office box address for general correspondence, and they are happy with the combination of flexibility and continuity that it gives them. Try to acquire your office equipment during the months before you open your practice, while you still have income from another source and while you have time to hunt for bargains. Just as you do not need an impressive office location, neither do you need fashionable office furniture and equipment. Used items, such as desks and typewriters, may be as serviceable as new ones, but will often cost less than half as much.

While bargain-hunting can save you a good deal of money, do not compromise on the quality of the typewriter you buy. Whether new or used, it should give a sharp, clear impression. In correspondence, the quality of your typewriter is a significant part of the image you will project.

You will need to have several items printed commercially. At the minimum, you will need business cards, a brochure, and stationery.

Although printing rates are usually lower for large quantities, it is probably better to order only a modest quantity of each item. Chances are that you will want to change one or more of them within a year or two. Do not have any of these items printed until you have completed the formalities of establishing your firm's name.

You can economize on printing costs by doing some comparison shopping before placing your orders. One leading consultant suggests seeking out printers who have a simple line listing in the yellow pages of the telephone book, rather than those who have large advertisements, since advertising costs are reflected in prices. Before placing your order, be sure to examine samples of the printer's work.

Financing the Startup. From this summary, you can see that it is possible to set up a consulting engineering practice with a minimal business investment. A more serious consideration is to provide for your personal and family expenses during the time before the practice becomes profitable.

One approach is to begin building your practice on a part-time basis, while you are still employed in a full-time day-time job. This would allow you to begin building a client base and establishing a reputation, while taking very little risk. You should not consider this approach, however, if your part-time work would put you in competition with your employer.

Even though you may start out consulting only part time, you will eventually want to move into full-time practice. When this occurs, no matter how successful you may have been while moonlighting, your personal income will take a sharp downward turn. It is essential to have sufficient funds to cushion this predictable downturn.

Almost all beginning consulting engineers must take these funds from savings, a spouse's income, or the investment of family members or friends. Under exceptional circumstances, you may be able to get a loan from a bank or other lending institution. Most banks, however, are reluctant to make loans when there is no building or inventory to serve as collateral. Likewise, most venture capitalists are looking for a share in the profits from the future sale of a tangible product. If you have a highly entrepreneurial bent, it is possible that you will be able to find external financing for the early days of your practice. It is more likely, however, that you will have to finance your startup from your own resources.

LICENSES For a consulting engineer, the most basic license is the state engineering license, awarded at the end of the registration process. This license is prerequisite to the independent practice of engineering. It is renewable upon payment of a fee at specified intervals, usually once a year. If you intend to practice in more than one state, you should obtain licenses in each of those states.

Almost every city and county in the U.S. requires businesses of all types to register and pay license fees in order to do business within its boundaries. You may obtain the necessary local licenses by applying and paying the required fees at the city hall and county offices.

You should also make sure that you do not violate city or county zoning ordinances in establishing a business in your home. If you are planning to do any building or remodeling, you must obtain the necessary building permits.

CHOOSING A NAME FOR YOUR FIRM In the midst of serious decisions about ownership, financing, accounting systems, and the rest, you will also need to choose a name for your consulting company. Most consulting engineers use their own names; for example, "John Doe, Consulting Engineer." Partners tend to use both names, as in "Smith & Jones Engineering Consultants." If there are too many partners to name, you may prefer something like "Smith, Jones & Associates." If you choose to incorporate, you should add "Corporation" or "Inc." at the end of the name.

Most states have rules regarding the use of names of individuals in the names of engineering companies. The main rules are that any individual who offers engineering services must be a registered engineer, and that any individual whose name is included in the firm's name must be a registered engineer. Registered architects, land surveyors, and geologists are also acceptable in most states. A firm can usually continue to use the name of a retired or deceased person in its name if certain conditions are met. You can use other types of names if you prefer. Some engineering firms have names like "XYZ Technology, Inc.," "Consulting Engineers Associates," or "Ourtown Engineering." You can use more imaginative names, but you should be sure that the name you choose does not make it difficult for a potential client to understand that you offer consulting engineering services.

FICTITIOUS NAMES   If your firm is a sole proprietorship or partnership, you must determine whether or not you want to use a fictitious name for your business. Names that do not include the surnames of the owner or of all general partners are considered to be fictitious names, as are names that imply the existence of additional owners (such as "Doe & Company," "Doe & Associates," "Smith-Jones Group," etc.). The name of a corporation, if used exactly as it is registered, is not considered to be a fictitious name. If you choose a fictitious name, you must file a fictitious business name statement and publish a notice that you intend to use the name. Check with the county clerk's office for the correct procedures.

Before you start using the name you have chosen, you must find out whether the name is already in use in your state and in the county or counties in which you intend to do business. It is best to have your attorney make these inquiries for you. If the name has been recorded or registered, you will need to alter the name or choose a new one.

EMPLOYEES, AND SOME ALTERNATIVES

The issue of whether to engage employees may be a thorny one for the beginning consultant. After your practice is well established and you have a steady volume of business, there is no doubt that you will want and need one or more employees. In the beginning, however, the decision to hire employees should be made according to your needs and budget.

Having someone to answer the telephone, type correspondence, and handle the daily bookkeeping has many advantages. It can free you from many time-consuming tasks, and give you more time to devote to your consulting work and marketing. Knowing that your telephone is covered, you will feel free to be away from the office.

On the other hand, having an employee costs a good deal of money: You will need to pay your employee and make contributions to Social Security, federal and state unemployment taxes, and employee benefits. There is a considerable burden of paperwork as well. In addition, you will spend time and energy on questions of supervision and management—just when the problems of establishing your practice are demanding most of your attention.

If you want the benefits of an employee but are not yet prepared to pay the costs, you should consider alternative ways of obtaining the services you need. Telephone answering typing, filing, and bookkeeping services can be available to you without hiring a secretary or office assistant.

Telephone coverage is extremely important. It is essential both for maintaining your continuous availability to potential clients and for projecting a professional image. If you cannot yet afford a secretary or receptionist, you must nevertheless arrange to have your telephone answered when you cannot be in the office.

If you work out of your home, you may prefer to have your spouse answer the telephone in your absence. (You should, of course, have a business telephone separate from your personal telephone.) Before making this choice, however, try to evaluate the image that you will project. Is your spouse willing to sound like a secretary? Does your spouse's schedule require being out of the house frequently? Are there small children whose voices may be heard in the background? Are there older children who may try to answer the telephone? Only if your spouse is always available and willing and able to adopt a professional telephone manner, should you consider this solution.

If you rent a desk or office space in a larger suite or office complex, telephone-answering services may be available for a modest fee. This is an excellent solution, if it fits your overall situation.

You may subscribe to a telephone-answering or call-forwarding service at reasonable rates. If you anticipate being out of your office a good deal of the time, this type of service will more than pay for itself by keeping your office "open" even when you are out.

The least desirable solution is a telephone answering machine. Despite the fact that these machines are now very common, they almost invariably leave a poor impression on the caller. Some callers may leave messages, but more will simply hang up. You are likely to lose a good deal of business by using an answering machine.

Correspondence that looks professionally typed is also very important to your business image. If you are a good typist and your volume of correspondence is not great, you may prefer to do the typing yourself, taking care to use only high-quality letterhead stationery and a good typewriter. If there is a competent typist among your adult family members, he or she may agree to serve as a part-time typist for you. As the volume of material to be typed increases, these solutions will probably become unworkable.

If you have very little typing to be done in the beginning, you may be able to manage for a while with a typist who is willing to work on a piece-work basis. The per-page cost may seem relatively high, but the total cost is much less than that of even a part-time employee.

Probably the most efficient solution is to hire a temporary employee through a temporary employment agency. You can hire a typist, or a person with multiple office skills, for as little or as much time as you need. For example, during slow periods, one day per week of general office assistance may be sufficient to maintain correspondence, other typing, filing, billing, and bookkeeping. On the other hand, if you are preparing several proposals at once, you may need full-time help for a couple of weeks. Working with a temporary agency gives you the flexibility to handle this kind of variation in work load. It also gives you the convenience of paying a flat fee, without worrying about taxes and paperwork that directly hiring an employee entails.

Like most other small businesses, your practice will probably not require the full-time services of a bookkeeper. In fact, bookkeeping is one task that you can safely (perhaps even profitably) reserve for yourself. If, however, it becomes too time-consuming or irksome, it is easy to purchase bookkeeping services. Some free lance bookkeepers work on an hourly or half- daily basis. Others prefer to work on a retainer basis, performing specified services for a monthly fee. In addition to bookkeeping services, you will probably want to retain an accountant to prepare tax statements for you.

INSURANCE

You should not go into business without adequate insurance coverage. The types and amounts of coverage you will need will depend on the field of engineering you intend to work in, whether you rent or own your office space, whether or not you have employees, the form of ownership of your business, and the balance between risk and the cost of insurance.

You will probably want to consider several different types of insurance. Some types of insurance that are most likely to be useful to you include the following:

General Liability Insurance. This is a must for almost every business. You are responsible for the physical safety of anyone who comes into your office, including employees.

Professional Liability Insurance. Design professionals, including consulting engineers, have with increasing frequency been found liable for various types of design and product failure. Professional liability insurance, or "errors and omissions" insurance, is an essential protection against such liability claims. This insurance should be chosen carefully, with an eye to deductible provisions.

Fire Insurance. If you own the building in which you operate your business, you will need fire insurance. The ordinary policy does not cover accounting records, bills, money, manuscripts, deeds, securities, or any other documents. Since these may be the most valuable items in your office, you may want additional coverage that includes them. If you rent your office, inquire about fire insurance. Your office may be covered, or you may need a separate policy for your furniture and belongings.

Comprehensive Insurance. You may want to consider comprehensive insurance against various types of crimes, such as burglary and forged checks.

Workers' Compensation Insurance. If you have any employees, workers' compensation insurance is mandatory.

Business Interruption Insurance. Insurance against business interruption can protect you against interruption or disruption of your business by unexpected events, such as fire, your illness, or some disaster befalling a contractor. Such insurance can be limited or comprehensive; you should negotiate the precise policy that you need.

Life Insurance. Aside from your personal life insurance policy, this will probably not concern you in the beginning. As your business grows and you acquire more employees, a group life insurance policy may be an important part of an employee benefits package.

Health Insurance. Except for your personal or family policy, health insurance, like life insurance, will probably not become an issue until you have employees, who will look upon health insurance as the most basic of fringe benefits.

Even this brief review shows that the question of insurance is a complicated one. Not only are there many different types of insurance, but for each type there are many different companies offering a variety of policies that differ in total coverage, deductibles, and premiums. You will almost certainly need some expert advice in order to choose wisely.

Unfortunately, it is not always easy to find a qualified expert whom you can trust. An insurance agent may have the necessary expertise, but, since an agent's fee varies with the size of the sale, he or she is likely to overestimate your need for coverage. Nevertheless, an insurance agent can provide you with a great deal of information. Try to find a good agent by asking friends and fellow professionals for their recommendations. Your lawyer or accountant may know of a good agent. Check to see whether the agent is a chartered underwriter (CPCU or CLU). This credential is a good indicator of competence and experience.

You should also do some comparison shopping. An insurance agent can offer you only the policies that are available through his or her company. Other insurance companies may offer coverage that is closer to your needs.

An alternative is to get the advice of an insurance consultant. A properly qualified insurance consultant will be a member of either the Institute of Risk Management Consultants or the Insurance Consultants Society and will not be an insurance agent or broker. Insurance consultants charge a fee comparable to that of other professionals, but your consultation will probably not last longer than an hour. If you get sound advice, you will save more than the consultant's fee in insurance premiums. Say no to offers of extra services by the consultant. You will not need them at this point.

MARKETING

If you are like most other beginning consulting engineers, the idea of marketing is apt to make you feel a bit uncomfortable. For one thing, you may feel that it is unprofessional. For another, you may be less prepared to handle this aspect of consulting than you are any other. Next to your engineering expertise, however, your marketing skills will have the most impact on the success of your practice. It is essential that you gain a perspective on marketing that allows you to do it skillfully and confidently.

WHAT IS MARKETING? Marketing is the sum total of activities that brings your name, professional capabilities, and availability to the attention of potential clients. It consists of all the methods you use to make potential clients aware of what you can do for them.

As such, marketing is much more than advertising. In fact, advertising may have little or no place in your marketing plan. Rather, marketing entails carefully planning and implementing a wide variety of activities, each of which is designed to attract potential clients.

DEFINING YOUR SERVICES The first step in developing an effective marketing plan is to define clearly and completely the services you will offer. Try to avoid the two mistakes most commonly made by beginning consultants. The first of these mistakes is to define the range of services so narrowly that you miss the opportunity to take on jobs that you could have done. The other is to define your range of services so broadly that you lose credibility or, worse, get involved in projects for which you are not prepared.

Try to imagine the actual work that you will be doing for clients as a consulting engineer. Do you see yourself taking on complete design projects in your field? Troubleshooting for clients? Writing proposals on behalf of clients? Offering advice on the selection of technology? Write a list of the tasks that you think you are most capable of doing and on which you want to spend most of your time. Then, try to think of the other skills that you have, and how you might apply them in the context of your consulting work. Make a list of these secondary tasks. These two lists will give you a preliminary definition of your capabilities as an engineering consultant.

If you have a great many skills and capabilities, it will be necessary for you to focus on the main areas of service that you want to offer. You will be presenting yourself as a professional with specialized skills. If you offer too many different kinds of service, a prospective client may doubt that you have expertise in all of them. Later, when your consulting practice has grown, you can add services without sacrificing credibility.

On the other hand, if you appear to specialize too narrowly, you may miss engagements that are well within your competence. If you are thinking of offering only one basic service, try to think of auxiliary services that you are capable of performing and that clients may need. You may even find that there is a better market for the auxiliary services than for the basic one. Although this may not be the case, you may want to take on other kinds of jobs when there is a lull between engagements of the type that you prefer. Later, when you have enough business to turn some jobs down, you can adhere more closely to your specialty.

DEFINING THE MARKET   The next step in developing a marketing plan is to define the market; that is, to decide who your potential clients are. Who is most likely to need, and buy, your services? Will your clients be builders? Manufacturers? Utility companies? Government agencies? Individuals or corporations? It is very important that you clearly and specifically identify your potential clients.

In defining your market, it may be helpful to prepare a written list of the types of individuals or organizations to which you expect to sell your services. Writing the list will help you focus on the characteristics of your potential clients and, at the same time, it may help you to think of new possibilities. Later, when you need lists of particular individuals and organizations, your list of categories will be a guide.

Another consideration is the location of potential clients. Do you expect to do all your business in or near the town or city in which your office will be located? Or, do you expect to draw clients from a larger geographical area? (If you expect to do business in more than one state, be sure that you are licensed as a professional engineer in each of the states you have in mind.) Do not ignore the possibility of international clients. The fact that your firm is small is no impediment: more than one-fourth of all engineering firms doing international business have five or fewer employees.

Try to estimate how many potential clients there are in the geographical area that you are considering. How often are they likely to need the types of service that you intend to offer? Is there a great enough need for those services to support both you and your competitors on a long-term basis? If the answers to any of these questions are discouraging, you may want to think about including a larger geographical area or expanding the scope of services that you intend to offer.

Having defined your market, at least in a general sense, your next task is to determine how to reach potential clients with the message about your capabilities and availability. In short, your next problem is how to handle advertising and public relations.

ADVERTISING   Your choice of methods for "business development" will be constrained by a number of factors, including professional ethics, legal regulations, and the relative effectiveness of the various methods.

Despite the competitive nature of the economy, direct competition among engineers has long been opposed by professional engineering associations. This has meant, among other things, that public advertising has been frowned upon as being unprofessional. In recent years, however, following Supreme Court rulings that restrictions on advertisement by professionals constitute restraint of trade, this stance has softened. As in the legal profession, some firms have begun publicly advertising their services.

Traditions die hard, however, and many professional engineers—and their clients—view public advertising with distaste, regarding it as undignified and unprofessional. Even though such advertising is no longer directly contrary to professional codes of ethics, it is likely to work to your disadvantage, since it may offend professional colleagues and undermine your professional image in the eyes of potential clients.

What methods, then, can the consulting engineer use to reach his potential clients?

The most effective form of advertising for a professional is word-of-mouth—that is, recommendations by satisfied clients. A reputation for competent performance of engineering work is the greatest marketing asset you can have. At the beginning of your consulting career, however, you will still be in the process of establishing your reputation. How can you get your message across to potential clients before your reputation is ready to speak for itself?

One basic marketing tool is your professional card. Your card should give your name, the name of your firm, your business address and telephone number, and possibly a brief indication of your specialty or the basic services you offer. Before having your card printed, study the cards of other consulting engineers. Select a format that will project a dignified, professional image. Carry a supply of cards with you wherever you go, and use them on every appropriate occasion. They will help others to remember your name later, when there is a need for the sort of engineering services that you offer.

The other basic marketing tool is a company brochure. You may, in fact, have several brochures, each designed for a different purpose. Your *capabilities brochure* will be most important. In this brochure you should list in detail the types of engineering services that you offer. The brochure should list all partners (if you are not working alone), and it should present information on the experience of all principals. If you have a supporting staff or special facilities, the brochure should include information on them, as well.

Your capabilities brochure should be fact-oriented, avoiding the use of laudatory terms or any kind of exaggeration. It should be designed to let the client know exactly what you can do and why you are qualified to do it. Puffery is common in product advertisements, but it has no place in professional-service advertising.

In addition to your capabilities brochure, you may want to produce one or more *sales brochures*. In a sales brochure, you may deviate from a strictly technical presentation of your capabilities. While self-laudatory and exaggerated claims are always out of place in professional advertising, it is legitimate and effective for a sales brochure to focus on the benefits you can bring to the client. That is, your sales brochure can make the potential client aware of problems, and then show how you can solve them.

There are many other types of sales aids, which generally are used with much less frequency than professional cards and brochures. Even though you may use them less frequently, it will be worthwhile to have them available when the occasion arises.

For example, you should maintain an up-to-date list of completed projects. For each entry you should include the dates, the client's name, and a brief description of the project. As this list grows in length, it can be an important aid to establishing your credibility in the eyes of a prospective client.

Similarly, you may wish to maintain a file of project descriptions that list the major features of the work you accomplished. These descriptions can be used to substantiate your experience with a particular type of project. Photographs of completed projects can also be convincing evidence of your ability and experience.

To demonstrate the quality of your work, you can show a client a sample of typical reports, plans, or specifications from previous projects. For clients who are primarily interested in having you prepare a plan or a report, such samples of your written work may carry more weight than descriptions or photographs of completed projects. (However, you must take care to observe ethical constraints on confidentiality and disclosure in presenting information prepared for others.)

If you have written professional articles, lists and reprints of them can contribute to your image as a serious and knowledgeable professional. In addition, writing articles on new or existing technology for engineering magazines is an excellent means of obtaining free, national advertising.

Similarly, you should speak at conventions, technical symposiums, and meetings of technical societies. Lists of such engagements, even if they were unpaid, serve to establish your image among potential clients. Making such presentations will help to establish your position among professional colleagues, as well.

You may be invited by a prospective client to give a presentation, either as a general introduction to your work or in conjunction with a proposal. The use of skillfully prepared audiovisual aids, such as slides, film strips, or videotapes illustrating some of your previous accomplishments, can be very effective.

Of course, all of these items—including cards and brochures—are merely aids. They must be used in an interpersonal context which you create. For the professional engineer, marketing is first and foremost a matter of personal contact.

PUBLIC RELATIONS

The activities that are collectively known as "networking" are essential to building and maintaining a successful consulting practice. The more you are personally known by potential clients, professional colleagues, and others who might recommend you for an engagement, the better chance you have of being chosen for any given job.

This means that you must allocate time—both business time and personal time—for cultivating the acquaintances and contacts that will ensure your success as a consultant.

Personal Calls. Personal calls on potential clients will be one of your basic marketing activities. You should keep updated lists of potential clients located in your geographical area, and you should call on each of them at their offices. Be prepared to learn about the client's need for engineering services and to explain briefly the services that you offer. You should be sure to leave your card and your brochure. If the potential client is a large organization, you may need to make several calls in order to become acquainted with all the people in the organization who are involved in the decision to procure engineering services.

Naturally, most of these calls will not lead immediately to engagements. They will, however, create an awareness in the mind of a potential client that your services are available. To maintain that awareness, it is advisable to keep up the contact through various means. Where appropriate, a telephone followup to a personal visit can be effective. Mail followups can also be useful. For instance, you can send specific information that the potential client may have asked for; or, you can send reprints of any of your writings that may be relevant.

After a period of time, another personal call may be in order. A followup visit helps to keep your name in the client's mind, but you should also use it to gain more information about the potential client's staff, engineering problems, and likelihood of future requirements for consulting engineering services. It is good to maintain periodic personal contact with all potential clients, so that you will be informed of upcoming projects.

As you call on each prospective client, your list of names will develop into a file-full of information that you can use in your future marketing activities. After every personal call on a prospective client, make a note of all relevant information. You should record the names and positions of the individuals you have talked with, as well as others in the organization who may be involved in decisions that would affect you. Keep a record also of present and anticipated engineering projects and problems. Before making a followup contact, use your file to review what you know about the client. This will help you to seem (and be) informed about, and interested in, the client's situation and needs.

In order to develop personal contacts effectively, you should plan ahead. Schedule your calls and followups strategically, using all the information that you can gather. Failure to plan can lead to disappointment and wasted effort.

Although making personal calls is quite time-consuming, you should not consider turning this task over to a sales person, even if your volume of business can support one. Your technical expertise, or that of a partner, is crucial in preliminary contacts with potential clients.

When a prospective client lets you know of a particular requirement for engineering services, you should act swiftly and strategically. You must now supply the client with your specific qualifications for performing the particular work that is required. You may be able to arrange for a presentation, or you may submit information and documents designed to demonstrate your competence to undertake the project.

If you are asked to make a presentation, it will usually require speaking to a group that has decision-making powers for the client. Depending on the nature of the client (corporation, municipal government, state agency, etc.), the technical competence of the group may vary. You may encounter unpredictable levels of technical expertise (e.g., in a city council or a board of directors), or you may encounter those with very strong engineering qualifications (e.g., an in-house engineering unit). Prepare your talk and your supporting materials carefully, bearing in mind the nature of the group. If the group includes people who are not knowledgeable in your field, you need to take special care to present the proposal and your qualifications in terms that can be understood by all. A well-delivered talk, supported by appropriate materials such as sample plans, reports, photographs, charts, and the like, can be a determining factor in obtaining an engagement.

News Releases and Announcements. Another way to keep your firm's name in the minds of potential clients is to make use of print and broadcast media to make noncommercial announcements. Such announcements may cover a variety of activities or developments. For example, the completion of a major project might be the occasion for a news release to local newspapers. The awarding of a new engagement or the beginning of a new project may also be announced as a news item.

Your activities, and those of your partners or other staff, may also be deemed newsworthy. For instance, if a new engineer joins your firm, you should announce it publicly. (In such a case, do not neglect to add the new person's name and qualifications to your brochure and other advertising aids.) If you participate in professional or civic activities, you can write this up as a news release as well.

Your local newspaper is probably the best place for publishing announcements of this type. Large metropolitan dailies may be less receptive than smaller community newspapers, which are usually eager for news of local business and civic activities. Scan your local paper for examples of the types of news items discussed here; you will find many that you can use as models.

Other media can be useful, also. A trade journal that is likely to be read by potential clients may be the ideal place to publish certain types of news items. Study the relevant trade journals to see what types of items might be appropriate. Newsletters in your area of specialization may be another outlet to consider. A newsletter editor will probably be willing to publish your news item if it fits the newsletter's general area of interest.

Local radio and television stations also offer opportunities for exposure. No matter what your specialty is, there is undoubtedly a local radio or television program that could provide you with a suitable forum.

Do not forget one of the most effective of all communication media: the personal appearance. An interesting speech or panel presentation at a trade or civic function can leave a favorable impression on potential clients.

Community Functions. Even if you do not give a speech or any other sort of presentation, regular participation in civic and community activities can increase your exposure and help to build your reputation. The people who are active in community and civic affairs are also very likely to be the decision makers in local businesses and local government. If these people know you personally from your community activities, they will think of you first when the services of a consulting engineer are needed.

It is also advisable to join one or more service or fraternal organizations. Many community leaders are active in organizations such as Lions Club, Rotary Club, or Masonic Lodge. Also, you can meet most of the local business leaders by joining the Chamber of Commerce in your city.

Professional Societies. Participation in the activities of professional societies can be an important source of referrals. There are two major types of societies to which you may wish to belong.

The first type is the technical society. This type of society is organized by engineers in a particular specialty area, such as civil engineering, mechanical engineering, or electrical engineering. (See Appendix A for a list of national engineering societies in various specialty areas.) Membership in one or more technical societies helps you to keep abreast of new technical developments, as well as to maintain contact with colleagues in your field. If you are active within the society, either in an organizational role or as a contributor of technical papers, you will increase the probability that your colleagues in the society will remember you and refer potential clients to you.

The second type of society is more oriented to the specific concerns of professional and consulting engineers. The National Society of Professional Engineers (NSPE), organized to promote the interests of professional engineers, has chapters in every state and is open to individual membership. One of its divisions, Professional Engineers in Private Practice (PEPP) is, as its name suggests, an association of consulting engineers. (See Appendix B for a list of state and regional affiliates of NSPE and PEPP.) The American Consulting Engineers Council (ACEC), a federation of state and regional member organizations, is exclusively devoted to the concerns of consulting engineers. Membership is open only to engineering firms.

Like membership in technical societies, participation in professional societies can provide you with both information and important contacts. It can also provide you with valuable support and services, from directory listings to professional development programs.

Referrals. In the long run, the most effective ingredient in your program of advertising and public relations will be client satisfaction. Every satisfactorily completed engagement creates the potential for a repeat engagement with the same client. Repeat performances are in themselves an excellent advertisement: they show that the client was sufficiently pleased with your work to award you another assignment. Even if the client has no further need for consulting engineering services and another engagement is not possible, a satisfied client is very likely to refer colleagues who need engineering services to you.

MARKET RESEARCH
To be effective, your marketing efforts must reach those individuals and organizations that have an actual need for your services. In the face of rapid technological development, the needs of clients may change over time. New services may become necessary, while some old services may become obsolete. The opportunities open to you, then, may shift as time passes.

In order to keep up with market trends, it is essential that you carry out market research on a continuing basis. This need not be a formal program while your consultancy is still small. You must be continually alert, however, to changes in technology and procedures in your field, and to changes in the nature of services required by clients.

TIMING
As long as you are in business as a consulting engineer, marketing will be one of your constant activities. Virtually no consulting firm arrives at the point where all business derives from referrals and repeat engagements. Even if you should be fortunate enough to receive many referrals or invitations to submit proposals, you must still convince each potential client that your firm is the

best for the job. It is best to recognize from the outset that marketing will be a never-ending activity.

Your initial marketing efforts may not produce immediate results. You may wait for weeks, months, or even longer before a particular program of contacts or other strategy for business development begins to pay off. Patience, determination, and a refusal to become discouraged may be required during periods when you receive no response from potential clients.

MARKETING ETHICS   As in every other aspect of your consulting career, you should carry out your marketing program in conformity with the highest standards of professional ethics. Some principles that you should constantly keep in mind include the following:

- Brochures and other forms of advertising should avoid exaggeration and self-laudatory claims. Be careful not to misrepresent your qualifications, your availability, or the competency of your staff.

- Never pay a fee or offer any sort of gift in order to obtain an engagement, and avoid any other improper practices in soliciting assignments.

- Remember that the reputation of the profession is just as important as your own reputation. Whatever you do in your professional capacity, including your marketing efforts, should reflect favorably on the consulting engineering profession.

- Claims about your own qualifications and competence should never be made at the expense of another consulting engineer's professional reputation.

SALES

The end product of successful marketing is a sale—that is, a client's agreement to purchase your engineering services. That agreement will usually be preceded by your submission of one or more proposals, as well as by negotiations concerning the scope of services and your fee. The agreement will usually be summarized in a written contract.

PROCUREMENT OF ENGINEERING SERVICES   Before discussing the details of proposals, fee-setting, contracts, and other matters concerning sales, it may be helpful to look at the procedures that clients use in selecting a consulting engineer.

In some instances, the client may choose a consulting engineer directly, without any formal selection procedures. The choice may be based on previous satisfactory experience with that consultant or on recommendations by a third party, or it may be the result of the consultant's marketing efforts. Most projects, however, are awarded, following a selection process, to one of several competing firms. The procedures used in the selection will vary according to the type of client (government agency or not), the scope of services required, and the preferences of the client.

The client ordinarily has two main concerns in evaluating consulting engineers for an engagement: quality of services and price. Most of the variations in selection procedures have to do with the relative weight given to these two concerns, and with the point at which price becomes a consideration.

In the traditional procedure, price becomes a matter of negotiation only after the client has ranked the competing firms according to their qualifications. The traditional procedure begins with the client preparing a list of three or more consulting firms considered to be qualified for the job. Each of these firms is then interviewed and given information on the services required. In all likelihood, each firm will be asked to submit qualifications proposals.

Following a thorough review of each consultant's qualifications, the client ranks the top three candidates and invites the first- choice firm to enter negotiations on fees and contract terms. If agreement can be reached, that firm is awarded the engagement. If agreement cannot be reached, the second-ranked firm is called in for fee negotiations. In the event of a second failure to agree, the third-ranked firm will be called upon.

The outstanding feature of the traditional method of selection is the importance given to the qualifications of the consulting engineer and to the quality of services expected. Before ranking the consulting firms, the client reviews all aspects of each consultant's experience, accomplishments, reputation, and understanding of the current problem. The consultant is interviewed closely, and the qualifications proposal is studied, before price is ever considered.

Most consulting engineers believe that this traditional procedure provides the best guarantee of high-quality service to the client. The U.S. government, whose various agencies collectively constitute perhaps the largest single market for consulting engineering services, appears to be in fundamental agreement with this view. Legislation regulating the procurement of design services (the "Brooks Bill" of 1972, or Public Law 92-582) requires that qualified firms be evaluated and ranked, and that the first-ranked firm be invited to negotiate fees and contract terms.

Despite a general resemblance to the traditional procedure, the procedures used by federal agencies are more complicated than those typically used by private clients. The requirement for services must be announced in a prescribed manner. (For instance, if the expected fee is over $10,000, the announcement must be published in the *Commerce Business Daily*.) Consultants who are interested in engagements with federal agencies usually must submit their qualifications and experience and any other information on standard forms. Depending on the agency and the nature of the services required, a variety of regulations regarding location, size, minority ownership, and other aspects of firms may apply to the selection of a consultant.

State and local governments and their agencies are not required to follow federal procurement procedures, but many of them do. Others follow the traditional procedure, and still others require one form or another of price bidding. Some private clients may also prefer some type of bidding procedure.

Price Bidding. One form of price bidding is the so-called "two-envelope system." In this system, the consulting firm submits two proposals at the same time. One envelope holds the qualifications and experience proposal, while the other contains a price proposal. Presumably, the client evaluates the qualifications of the competing consultants and ranks them before examining the price proposals. There is no guarantee, however, that this sequence will be followed.

In another variation of price competition, a few consultants are selected on the basis of qualifications and are then asked to submit price proposals. On the presumption that each of the bidders is adequately qualified, the engagement is awarded on the basis of the price proposal.

Direct price bidding is another alternative, but few clients are willing to take such a risk in awarding a design assignment. Although direct price bidding may be an appropriate means for selecting some types of contractors and suppliers, experience and appropriate qualifications are vital in design work. A consultant can usually cut costs only by devoting less time, thought, imagination, and care to a project. Recognizing that competent professional services cannot be fully quantified, most clients prefer a method of selection that gives qualifications priority over price.

PROPOSALS   One of the fundamental tasks of a consulting engineer is the preparation of proposals. Proposals may be looked upon as the intermediate step between marketing and sales. If a potential client has decided that you are a good candidate for a particular engagement, you will probably be asked to submit a proposal. Depending on the selection procedure the client uses, you may be asked to submit a qualifications proposal, a conceptual proposal, a fee proposal, or some combination of these.

Qualifications and Experience Proposal. The type of proposal that you will most frequently prepare is the qualifications and experience (Q & E) proposal. The information included in a qualifications proposal is essentially the same as that contained in the capabilities brochure, but it is more detailed and is oriented to the particular engagement for which you are being considered. You may wish to include information on the history of your firm, emphasizing experience on projects similar to the one in question. Descriptions and photographs of previously completed projects may be included. You should give biographical sketches of yourself and any other principals. If you propose to associate with another firm for the purpose of carrying out the project, include the same sort of information about the other firm. Draw clear connections between your qualifications, experience, and facilities and the needs of the client.

Conceptual Proposal. A second type of proposal is the conceptual proposal. In the conceptual proposal you will directly address the problem faced by the client, presenting your view of what needs to be done and proposing your solutions. Preparing a conceptual proposal will take a considerable amount of your time and effort, but you may be well-rewarded by receiving the engagement.

While some clients will request a conceptual proposal along with a Q & E proposal, others will not. Even if it is not requested, if you have sufficient information about the client's needs to produce a good proposal, you may wish to prepare and submit a conceptual proposal. If you can show that you have a solid grasp of the client's needs and the constraints under which the work must be done, and if you can propose an effective solution, you will have gone far toward persuading the client that you are the best-qualified candidate for the job.

Fee Proposal. The third type of proposal is the fee proposal. The form and content of a fee proposal are dictated by the size of the project, the client's requirements, and the basis of the fee (lump sum, time rates, salary, retainer, cost plus percentage, or one of their variations). The fee proposal may or may not include a suggested contract. If a suggested contract is included, the fee proposal may be included as a part of the contract. If the client so desires, a detailed cost analysis may be submitted as part of the proposal.

SETTING FEES   In the past, professional societies published fee-setting guidelines which provided the consulting engineer (and the client) with information on widely-accepted methods of calculating fees. However, following court decisions in the 1970's that held that such guidelines constituted restraint of trade, professional societies were forced to give up formulating and publishing recommendations regarding fees. Today, each consulting engineering firm must develop its own method of fee setting.

In setting fees, you must strike a rather delicate balance: The fee you charge must assure you of receiving adequate compensation for your time, skill, and effort; at the same time, it must ensure that the client is receiving good value for his money. A fee that accomplishes both of these will be regarded as fair by both client and consultant.

Remember that selection is rarely made on the basis of price alone. Therefore, being the low bidder in a competitive situations will not necessarily ensure your being awarded the engagement. The most important consideration in making a bid is the setting of a fair fee—one that allows you to do a good job, be adequately compensated, and assure the client of good value.

Costs. From your point of view, the fee you charge for an engagement should be large enough to cover your expenses (costs) and to provide you with a reasonable profit margin. Your expenses, or costs, will fall into two basic categories: direct costs and indirect costs.

Direct costs are all expenses that are directly attributable to a particular engagement. Direct costs include items such as the amount of time you spend on a project (your salary), the time spent on the project by any other professional members of your staff, transportation, communication, reprographics, laboratory analyses, and any other expenses that are identifiable as project-related.

Do not try to recover direct costs from another project or from a period of activity on another proposal by charging more for direct costs on a current proposal. If your bid is too high, you will not get the job. Until your business becomes well established, your salary and other expenses will have to come out of your start-up capital.

Indirect costs are also sometimes called overhead costs or general and administrative (G & A) costs. These costs include all the costs of keeping yourself in business as a consultant—rent, utilities, parking, insurance, support staff, supplies, marketing expenses, proposal preparation, subscriptions, memberships, and the like. You will include a prorated share of your indirect costs in the fee for each engagement.

Rather than trying to calculate the precise amount of indirect costs that should be included in a particular fee, it is more practical to include a fraction of your direct costs as the indirect cost component of your fee.

After you have been in business for a year or two, it will be fairly simple to determine how much to charge for indirect costs. The basic method is to calculate the ratio between the total of your indirect costs and the total of your direct costs over a year's time, and to add that fraction to your direct costs.

For example, suppose that the direct costs for all the work you did last year came to a total of $83,000. Suppose also that your total overhead or indirect costs for the year were $74,000. This means that your indirect costs were 89 percent of your direct costs.

Assuming that this ratio will continue to be approximately the same in the current year, you should add 89 percent to your estimated direct costs for any given project in order to cover indirect costs. For example, if you estimate that your salary and other direct costs for a particular project will be $5,000, you should add 89 percent or $4450, so that the total cost component of the fee would be $9,450. (Note that, since the profit margin has not been added yet, this is not the full fee; it is only the cost component.)

In your first year of operation, it will be more difficult to figure the proportion of indirect to direct costs. You should be able to estimate your overhead costs fairly accurately, but it is difficult to know how much business you will do and, therefore, how much to expect in the way of direct costs. You will have to make an estimate, possibly a fairly arbitrary one.

For example, you may assume arbitrarily that you will have enough business to spend about 20 hours per week on billable work. Assume also that you have decided to accept a salary of $20,000 for the first year, or until you have enough business to give yourself a raise. This means that your direct salary cost will come to about $20 for each hour that you expect to spend on a project. To this you will need to add other anticipated direct costs. Depending on your specialty, the size of your geographic area, and other factors, travel, reprographics, and other items may raise your direct costs considerably.

For example, assume that your estimate for total direct costs in this situation is $27,000, and your estimate for indirect costs is $23,000.

Your indirect cost rate is 85 percent. For every project in your first year, you would add 85 percent of the estimated direct costs for each project in order to calculate the cost component of the fee for that project.

You should calculate your indirect cost/direct cost ratio at the end of each year of operation, to be sure that you are adding in a currently accurate percentage.

It should be clear from these examples that high overhead costs will result in accordingly high fees. You should try to hold your indirect costs to a level that will allow you to set reasonably competitive fees.

If you have a number of ongoing projects, you may find that your anticipated overhead costs for the year have already been covered. In such a situation you may be tempted to lower your proposed fees in bids on new projects by excluding indirect costs. Resist the temptation; it is important to maintain a consistent fee structure.

Remember that it is normal not to get every job you bid on. If you are getting nearly every job, you are setting your fees too low. If you never get any jobs, then your fees are almost certainly too high. You should continue to adjust your rates until you are consistently receiving a reasonable fraction of the jobs that you bid on.

Profit. In order to calculate your full fee, you must add a percentage of the cost component as profit. There is no formula for determining your profit margin, as there is for determining indirect costs. You must adjust your profit margin so that your fee will not be so high or so low as to put you out of the normal range for the market. If your fees are too high to be competitive, it is obvious that you will lose business. In the short run you may remedy this situation by keeping your profit margin low. You will be more successful, however, by reducing your indirect costs and maintaining a reasonable profit percentage.

It may be less obvious that excessively low fees can make you lose business by damaging your credibility. If your fees appear to be far below the average for your market, you should raise them to a level that is competitive and not unusually low.

It is very important to remember that profit is not the same thing as your salary. Your salary for the time that you devote to a given project is one of the direct costs of the project. Profit is income to your business over and above direct and indirect expenses. If you are a sole proprietor or partner, the tax system will not differentiate between your salary and your profits, but you must make the distinction clearly for yourself in order to set your fees correctly. If your firm is a corporation, it will be easier to keep profit and your salary separated.

In setting the level of your profit margin, remember that taxes will be paid out of profits. Although the exact proportion may vary a good deal according to the form of ownership and the size of your business, you can probably count on about half of your profits going to taxes. If you wish to retain five percent of your income as profit, your pretax profit margin should be approximately 10 percent.

Types of Fee Structures. Consulting engineering firms use a variety of fee structures. The main categories include the following:

- Retainer
- Lump sum
- Cost plus fixed fee
- Per diem method
- Salary plus percentage plus expenses
- Percentage of the total construction cost

Each of these methods of setting fees may have advantages or disadvantages for you, depending on your individual situation. Each of them requires careful cost accounting and clear communication with the client.

A *retainer* may be appropriate when a client needs consulting engineering services on a continuing or recurring basis. In such a case, you may agree to provide these services on call in exchange for a set fee. This fee may be paid as a one-time lump sum or it may be paid on a monthly or yearly basis.

The amount of the retainer is calculated by estimating the amount of time you expect to devote to the client. The direct and indirect costs associated with that amount of time, plus your profit margin, will determine the amount of the retainer. Your agreement with the client may include a provision for additional reimbursement on some specified basis if the time required exceeds a certain limit. It is customary to specify some costs as directly reimbursable by the client.

A retainer may also be used as an advance payment to be deducted from the final bill. This may be desirable in certain situations, such as when the final amount of the fee cannot be estimated accurately ahead of time, but initial costs are expected to be high. More frequently, a retainer may function as a guarantee that the client will actually proceed with the proposed project. If the client is unknown to you, or if you have any doubt about the client's ability or willingness to pay your fee, it is advisable to require a retainer amounting to approximately one-third of the total estimated fee.

The *lump sum* or fixed fee structure is most suitable for relatively small engagements in which the required services are clearly specified. If you make an agreement on a lump sum basis, you will be bound to the agreed-upon figure even if the project turns out to be more difficult than you originally anticipated. Therefore, it is best to be cautious in committing yourself to a lump sum fee, although some clients may insist on this fee structure as a protection against unpredicted expenses.

Like any other type of fee, a lump sum fee is calculated on the basis of anticipated direct and indirect costs plus profit. It should also include an allowance to cover unexpected events outside your control. You may also include an allowance to cover any unusual risks involved in the engagement. You may agree with the client that certain specified expenses will be reimbursed directly and not included in the fee.

The *cost plus fixed fee* structure is actually a variant of the lump sum fee method. The main difference is that, in the cost plus fee method, direct and indirect costs are itemized and billed as such, while the fixed fee is basically the profit margin. The fixed fee must be set so that it covers taxes, return on investment, compensation for risks, and any expenses or costs that may have been excluded from the reimbursable cost category.

With the *per diem*—time plus expenses—method of payment, you charge the client for the time you spend on an engagement at a particular rate per day, half-day, or hour, plus reimbursement for specified direct expenses.

If you wish to bill on this basis, the main problem is to set your rate so that it will cover your direct and indirect expenses plus profit. If you are working alone, you can use the basic method discussed in the Costs section. However, since direct costs other than your salary will probably be included in the category of reimbursable expenses, you may need a higher multiplier in order to recover your indirect costs. After adding your profit margin, your per diem rate will probably be something like two and one-half or three times your actual salary.

If you employ other engineers, assistants, technicians, etc., their per diem rates will be calculated in a similar manner but will be proportionately lower than yours. Partners, of course, should have equal rates. For some particularly demanding engagements, such as serving as an expert witness, your per diem rate may be somewhat higher than usual.

The *salary plus* percentage plus expenses structure is fundamentally the same as the per diem or time plus expenses structure. The main difference is the way in which the cost breakdown is made for the client. With the time plus fee method, the basic unit is the per diem fee. In the salary plus fee method, the basic unit is the direct salary received by you or your employees. To this you add a multiplier to cover indirect costs and profit, and the client reimburses specified direct costs.

If you are providing comprehensive engineering or design services on a project that involves construction, you or the client may prefer to set your fee as *some percentage of the total construction costs.* Normally, some costs (e.g., cost of land, legal costs, etc.) are not included in the total, and some engineering services (plan revisions, value engineering, management and supervision services, etc.) are billed separately. This type of fee is most commonly used for very large construction projects, and it is unlikely that you will undertake such projects until your firm has grown considerably. Therefore, you will probably not encounter the need for this type of fee setting during your first few years in business.

NEGOTIATIONS AND CONTRACTS

NEGOTIATING FEES   A client who is using the traditional procedure for selection will call you in for negotiation of fees and contract. If the client is a government agency, you may be invited to revise your fee proposal and to make a "best and final" offer. This offer will then be the basis for any further negotiations that may take place. A client may be favorably impressed with your proposal but still may wish to reduce costs or alter certain items of service. In all of these cases, your negotiations with the client must be face-to-face.

The most important point to remember is that, no matter how much the client likes your proposal, until you reach an agreement, the sale will not have been made. In addition to the careful cost accounting which is the basis for your fee proposals, you will need skill in the art of listening and persuasion in order to be successful in negotiations.

Some clients, after reviewing your original offer, will make it clear that they would like to have costs reduced. In such a case, you must review your proposal and determine which items of service can be sacrificed without interfering with accomplishment of the main goal. Under no circumstances should you reduce your price without cutting out some service. Otherwise, it would appear that your original proposal was overpriced, thereby reducing the client's confidence in you.

Many clients, despite their sensitivity to costs, are unwilling to sacrifice any services. When you make it plain that a cut in price entails a reduction in the level of service, the client may agree to your original price. Others may be quite willing to forgo certain items of service in order to cut costs.

Some clients may be reluctant to admit that they are trying to cut costs. Such clients attempts may negotiate a reduction in price by suggesting that you cut out certain items of service or by criticizing your approach to the project. With such a client, it is best to be tactful about money matters, and to guide the discussion toward the essential objectives of the project. If elimination of a particular service would hinder accomplishment of the main goal, this should be made clear to the client. It may be possible to cut out another item, which is less vitally related to the primary goal, instead.

Another consideration in negotiations is the client's view about what needs to be done and how it needs to be done. Some clients may be design professionals themselves, or they may have the advice of an engineering unit within their organization. Such clients may have strong opinions about how the project should be approached. Even non-professional clients may have their own ideas about the methods you should use.

It is best not to argue with a client's opinions. The most likely result of arguing is that the client will become more attached to his or her position. You should listen carefully to the client's ideas and show that you appreciate their value. If possible, postpone discussion of the points in question to a later time, or try to create opportunities for the client to change his or her mind without "losing face." Do not overlook the possibility that some of the client's ideas may be useful.

Whatever the substance of the discussions during negotiation, your first priority should be to understand the concerns of the client. Pay close attention to what the client says. Nonverbal signs, such as tone of voice, posture, and facial expression, can give you clues about what makes the client feel uncomfortable, hesitant, or reluctant. Sometimes the points that the client brings up may only camouflage other issues (such as price) which may be more embarrassing. As a successful negotiator, you must be alert to signals of concern and handle sensitive topics with tact. When you can satisfy the concerns of the client and, at the same time, be satisfied with the approach to the project and the fee, you are ready to make the agreement or contract.

CONTRACTS

The word *contract* usually brings to mind the image of a document filled with indecipherable legal terminology and "fine print." At the same time, a contract seems to promise the security of "getting it in writing," protecting both parties from the hazards of verbal misunderstandings.

In actuality, a contract is simply an agreement. A written contract is merely the documentation of the real contract, which is the agreement. Many of your engagements will not require such written documentation. For instance, small engagements with clients that you know well can often proceed on the basis of verbal agreement, provided that the scope of services and your fee have been adequately discussed. However, if the scope of services is great, or if you are not well acquainted with the client, it is advisable for you to document your business agreements with written contracts.

Your written contracts need not contain intimidating legal language. All types of contracts can be written either in the traditional style or in more modern English.

For relatively small projects, a simple letter of agreement will usually suffice. For larger, more complex projects, both you and the client may prefer a more formal document. Some clients may prefer to use a purchase order, which, with suitable modifications, can function as a contract. Be aware, however, that the terms of the purchase order (which are frequently written on the back of the order) take precedence over any conflicting terms that took place during your verbal agreement. Differences in terms and conditions should be worked out prior to the issuance of the purchase order and your acceptance of it.

No matter what their formats may be, all contracts must contain certain basic elements. The most important of these are the following:

- The identity of the contracting parties
- The competence or authority of the signatories to enter into a binding contract
- A commitment by the client to engage the consultant for services in a specified project
- A detailed description of the services to be performed
- Fees
- Terms of payment
- Duties and obligations of both parties
- Protection for both parties
- Effective date of agreement
- Date of expiration of agreement
- Standard "boilerplate" clauses covering unavoidable delays, unenforceable clauses, indemnification and warranties, and assignment of rights and responsibilities
- Signatures of both parties

Formal Contracts. A formal contract is a separate document prepared expressly for the purpose of describing and attesting to an agreement. It may conform to one of several relatively standard formats.

A formal contract begins by identifying the contracting parties by full name and address. Each party may be referred to in the body of the contract by a shorter term. For instance, your identification in the contract may read, "John Doe, Consulting Engineer, 400 W. Broadway, Ourtown, hereinafter to be referred to as Consultant," while the client may be identified as "XYZ Products, Inc., 750 N. Main Street, Ourtown, hereinafter to be referred to as Client." Other shorthand terms may be chosen if the parties so desire. The identification of the parties must include both the company's name and the individual's name, if they are different, as well as a specification of the legal form of the entity entering the agreement (individual, sole proprietorship, general partnership, corporation, etc.).

The beginning of the contract should also include the effective date of the agreement. This is particularly important if the effective date is different from the date on which the contract is signed.

Most formal contracts then proceed to a section which is usually called "Recitals" or "Recitations." Old-fashioned contracts may use the word "WITNESSETH" to begin this section, while modern ones may use the word "Background." In traditional contract language, most statements in the recitals begin with the word "WHEREAS." For this reason, this section is sometimes called the "whereas clause." Regardless of the wording, the main point of the recitals is to establish the authority of the parties to enter into a contract and to give other information pertinent to the validity of the contract.

The substantive part of a formal contract is often introduced by the words "NOW, THEREFORE." This section describes the project, commits the client to retain the consultant for services on that project, and describes the nature and scope of services to be performed. Each item of service that is to be provided should be identified and described in some detail. If the client intends to provide certain services (for example, surveys, laboratory tests, geotechnical investigations, or the like), these should be specifically excluded from the scope of services. This section should also include clauses designed to protect both parties in cases of amendment or termination of the contract or in cases of emergency situations outside the control of one or both parties.

The final section of a formal contract generally reads, "IN WITNESS WHEREOF, this agreement has been executed on (date)," and is followed by the names of the contracting parties and their signatures. If the parties are organizations rather than individuals (as will usually be the case), the names of the organizations, as well as the names and organizational titles of the persons signing the contract, must be included.

A formal contract can be prepared by a lawyer specifically for a particular engagement. This may be advisable for exceptionally large projects or projects which are unusual in some other way.

You may prefer to develop a standard contract for use with most clients. There are several advantages to having a standard contract form. Since only the details regarding a given project need to be typed in, a standard form simplifies the process of preparing a contract. When using a standard form, you are less likely to forget to include a necessary clause. Also, you can have the contract form drafted in such a way as to build in the particular provisions needed for adequate protection in your field of specialty.

Some professional engineering societies (ACEC, NSPE's PEPP, and ASCE) have formed a joint committee to make recommendations on contracts and other documents that are frequently used by consulting engineers. This joint committee (Engineers' Joint Contract Documents Committee, or EJCDC) has published suggestions for various forms of contracts, including both general contracts for engineering services and contracts geared to construction projects. You should study these suggested documents, available through the participating societies, before developing your own standard contract form.

It is advisable to obtain the assistance of an attorney in preparing a standard contract form. Even after studying the EJCDC suggestions and other examples, you will probably need legal advice in drawing up the contract form that will suit your needs best.

Letter Agreements. Letter agreements are frequently used in consultancy agreements, especially when the engagements are relatively small. As the name suggests, this type of contract takes the form of a letter. Because of this form and the more straightforward language that is generally used in a letter agreement, it seems to be a more informal document than a formal contract. In fact, a letter agreement must contain the same basic information as a formal contract and, when signed by both parties, has the same legal force.

In a letter agreement, one of the contracting parties, the sender, is identified by the letterhead on which the letter is typed, and the other party is identified by the name and address of the recipient. Therefore, there is no need for a separate paragraph naming the contracting parties.

After the conventional salutation (Dear Mr. Doe:), a letter agreement begins with a reference to the verbal agreement reached during negotiations. If there has been no face-to-face negotiation, the reference will be to previous correspondence, including any proposals you may have submitted.

Following this introduction, the letter will proceed to specify the services to be performed, the fee and terms of payment, the effective date of the contract, and the expiration date, if any. It may also specify any services that are to be excluded, obligations of both parties, and provisions for protection of both parties.

The letter will continue with a statement to the effect that acceptance will constitute a legally binding agreement and a request that, if accepted, a signed copy of the letter be returned to the sender. This is followed by the closing, the signature of the sender, and spaces for the recipient's signature and the date. Normally, two copies are sent, so that one can be kept and the other returned to the sender.

As with formal contracts, it may be desirable to prepare a standardized form for letter agreements. Although it is preferable to have each agreement typed individually on your letterhead, it is helpful to have a standard form to refer to, so that you will remember to include all necessary provisions. EJCDC has produced a suggested letter agreement form that may help you to draft your own form. As with standard contracts, it is advisable to consult a lawyer for the basic format of a letter agreement.

A letter agreement may, of course, originate with the client rather than with you. In this case, study the agreement carefully before signing.

Purchase Orders. Some clients prefer to use a purchase order rather than a formal contract or letter agreement. This is because their purchasing procedures are set up to deal exclusively with purchase orders. From your point of view, this should create no difficulties, so long as the purchase order contains a statement of the conditions of agreement, or such a statement is attached to the order. Check the general provisions, which may be printed on the back of the order, to be sure that they are acceptable to you. In particular, consider the implications of clauses that state that all terms of the purchase order will be deemed to have been accepted if the work is performed, regardless of whether the purchase order is formally accepted in writing.

Other Contract Forms. When you make a bid or offer on a job for a federal agency, the form you fill out may include a space for award of the project. Since you will have signed the form in submitting your offer, the signature of an authorized official in the "Award" section of the form transforms it from an offer into a contract. In making an offer on such a form, be especially careful in checking the accuracy of your figures and the description of the scope of services. If the agency accepts your offer, you are committed to perform the specified services for the specified fee.

State and local government agencies may use similar forms. Before submitting a bid on standard forms to any organization, check to see whether or not the form is a potential contract.

Particularly when working on small projects, some clients shy away from written contracts. If you have come to an adequate mutual understanding regarding the scope of services, fees, and terms, it is certainly possible to proceed without a written contract. In some cases, however, you may wish to send a letter to the client, expressing your appreciation for the engagement and confirming your agreement. You may ask the client to notify you immediately if there is any discrepancy between your letter and the oral agreement. Such a letter does not constitute a contract, but it may serve to minimize misunderstandings.

Special Considerations. Depending on the nature of the engagement, you, the client, or both, may have special obligations. Either or both of you may have special needs for protection, as well. For example, on international assignments, the client should arrange for visas, government permission, and assistance with customs. Or, if the client possesses records, plans, specifications, or other materials relevant to the project, he may be responsible for making these available to you. The client may require some form of guarantee that the work will be completed on schedule, or that it will not be subcontracted without written consent. You may be required to show proof that you carry certain kinds of insurance. All applicable conditions, such as those mentioned above, should be included in any agreement.

Understanding the Contract. An agreement or contract is binding on both parties. That is, either party can, through the courts, force the other to live up to the obligations of the contract. Legal redress, however, is seldom as satisfactory as is acceptable performance in the first instance. Since one of the main sources of dissatisfaction in a project is differing interpretations of the agreement, your prime consideration in generating a contract is to establish a clear understanding between you and the client.

If there is a written contract, adequate mutual understanding depends partly on the clarity of the document. All relevant details of dates, duties, obligations, scope of services, fees, terms of payment, and special considerations should be clearly spelled out in the contract. The language should be as straightforward and unambiguous as possible.

However, a written contract, no matter how well written, can never substitute for thorough discussion of the terms of the agreement. The time devoted to such discussion is time well invested. Points of misunderstanding or disagreement can be dealt with in the beginning, preventing disappointment and dissatisfaction later. Before signing a contract or otherwise sealing an agreement, it is always advisable to talk over the agreement until you are sure that both you and the client have the same understanding of and expectations from the project.

CLIENT RELATIONS

Your relationship with a client begins at the moment of first contact, whether through your marketing efforts, reference by a mutual acquaintance, or the initiative of the client. It will then continue through some or all of the phases of proposal, negotiation, agreement, performance, completion, and repeat engagement. Each of these phases brings its unique challenges to your management of the relationship. On the other hand, some basic principles apply equally to all phases of the consultant-client relationship.

Your first contact with the client will tend to set the tone for the relationship. For this reason you should endeavor, in all your advertisements, personal encounters, and public presentations, to maintain the image of a competent professional.

Your brochures and other printed materials should concentrate on the facts of your abilities and accomplishments. In meeting potential clients or making presentations, present yourself in a dignified manner. By dressing conservatively, you will probably make a good first impression. Equally important is your style of talking. Speak with self-confidence, but do not try to impress or overwhelm the client by talking quickly or loudly. Even if your client uses profanity, do not use it in your own conversation: Profanity always makes a bad impression. Avoid items that might offend, such as racy or ethnic jokes. In fact, it is probably better not to tell jokes at all. Your job is to perform professional services, not to entertain.

QUALIFYING CLIENTS   When you pass beyond the initial marketing stage and begin to discuss the possibility of a particular project, you should "qualify" the client. That is, you should gather information that will let you know whether you are progressing toward a possible engagement, or whether you are simply wasting your time.

One important piece of information is whether the person you are dealing with has the authority to reach an agreement with a consultant. If you are talking with a sole proprietor, a partner, a president, or a clearly authorized company officer, there will be no doubt about that person's authority to enter a contract with you. It is not so clear, however, when you are dealing with officials in government agencies or other bureaucratic organizations. In such a case, it is best to inquire diplomatically who is in charge of procurement of consulting services.

Another crucial matter is that of money: Does the client have both the ability and the intention to pay for the services under discussion? In some cases, particularly with government agencies or large organizations that are investigating the feasibility of an advance project, funds for such a project may not yet have been budgeted. In these cases you can probably get accurate information about the financial picture with a few tactful questions. For instance, you can inquire about the anticipated time frame (e.g., does the client expect to commission the project in the current fiscal year, or wait until next year?). You should also get information about billing procedures in such organizations, since they may be quite different from procedures that you are familiar with.

In other cases, you may have doubts about the client's seriousness, or about his or her ability to pay. Do not forget that appearances can be deceiving; you should not rely on your impression of the client's office for an estimate of his or her financial status. Be alert to potential trouble when a client seems reluctant to sign a contract, preferring a "gentleman's agreement." Other danger signals are a seeming lack of interest in financial details or a complete absence of negotiation and bargaining.

In such cases, you should bring up the subject of a retainer in the amount of approximately one-third of your total fee. In general, it is good policy to require a retainer, so that you are not in the position of extending large amounts of credit to the client. However, when there is doubt about the client's financial status or intentions, a retainer is almost mandatory. Likewise, a written contract or letter of agreement should be a routine part of every assignment, but it is especially important if you are in doubt about the client's willingness or ability to pay.

When the services in question consist primarily of analysis and advice, you must be especially careful to define the beginning of your billable time. Some clients may try to "pick your brain," get your opinions, and then decide that they do not need a consultant after all. To avoid such a situation, you should make it clear that the time you devote to discussing and analyzing the client's problem is billable time—even if it is the first meeting. Otherwise, you may find that you have given away the very services that you are in business to sell.

COMMUNICATION AND COOPERATION During the phases preceding the agreement, your first priority should be clear communication. At every point, you should attempt to understand the client's definition of the problem, the organizational, temporal, financial, personnel, and resource constraints under which the project will be carried out, and the client's worries and concerns. This will require careful and active listening on your part, as well as feedback from you to the client showing that you have fully grasped the problem. You should also make your own requirements and constraints clear to the client. Clear communication will pave the way for a satisfactory agreement or contract.

When the agreement is made, be sure that every relevant item is included in the written document and that each item is discussed until both sides have the same interpretation of their respective rights and obligations. Disagreements are especially likely to occur with regard to fees and terms of payment. If direct costs are to be billed to the client, billable items should be specified. The basis of payments (monthly, upon completion of a specified percentage of the work, etc.) should be agreed upon. If your contract consists of a modified purchase order, it may specify that payment will be made upon completion of your services. Such a provision may make it very difficult to obtain interim payments. If such a provision is unsatisfactory to you, you should discuss it with the client before signing the contract. If progress reports are required, their form and content should be specified. Final dates for full payment should be indicated.

While you are carrying out your work, you will need to keep the client informed about your progress. You should identify one person within the client organization as your liaison, and maintain regular communication with that person. In some projects, virtually constant communication may be required. In others, regular progress reports, at agreed-upon intervals, may be sufficient.

Cooperation between you and the client is essential for the smooth accomplishment of your mutual goal. The nature of the cooperation may vary widely from project to project—it may range from basic matters such as handling correspondence promptly, to supplying necessary information and documents, to assigning extra personnel to work on the project. Whatever kind of assistance may be needed by either side, a cooperative attitude and confidence in the other party's willingness to cooperate make for a productive relationship and a successful engagement.

AMENDING AND ENDING AGREEMENTS   During a project, either you or the client may feel the need to amend your agreement in some way. In such a case, it is essential to discuss the proposed changes thoroughly, and it is highly desirable to document these changes with a signed amendment to your original contract. If the changes are substantial, it may be preferable to negotiate a new contract to replace the original one.

The completion of the project is not necessarily the end of the relationship with the client. If all has gone well—if the client is satisfied with your work, and if the client has met his obligations to you satisfactorily—the basis has been laid for a long-lasting relationship which may result in referrals or repeat engagements, and will provide personal and professional fulfillment for you and your client.

PROBLEMS WITH CLIENTS   Unfortunately, not all engagements end so happily. Occasionally, a client will be unhappy with your work. On the other hand, you may be dissatisfied with a client who fails to pay your fee on time or who creates other troubles.

When bills are not paid on time, and when normal monthly billing procedures fail to produce results, it may be necessary to send reminders that are progressively less diplomatic in tone. If such reminders are also unsuccessful, you may have to threaten to turn the account over to a collection agency or to take legal action. As a last resort, you may have to implement the threats.

For smaller amounts, you can use small claims court to try to collect the amount due you. Where larger amounts are in question, you will need help in collecting, or you will have to file a suit against the client. Unfortunately, you may have to settle for less than complete satisfaction in such cases. A collection agency or lawyer who collects for you on a contingency fee basis is likely to keep a large percentage of the money collected. To obtain a court decision in your favor may cost you considerable time and money, and you may find it difficult, if not impossible, to enforce the judgment.

In the worst case, you may have to write off the loss as a bad debt, and include bad debts as an element in your overhead or indirect costs.

Bad debts not only result in direct financial loss, but they also lead to unnecessarily high fees and consequent difficulty in finding engagements. They can cause you personal worry and stress as well. Obviously, bad debts are to be avoided if at all possible.

Client dissatisfaction has almost equally unpleasant consequences. Not only does it mark the end of your professional association with that client, but it may also result in loss of future business, if the client fails to refer colleagues to you or even actively works against you. Remember that you have legal recourse in cases of defamation through libel and slander. Such cases should be pursued aggressively to maintain your reputation.

While it may be impossible to prevent entirely an occasional bad debt or a hopeless misunderstanding with a client, it is possible to reduce such situations to a minimum. Thorough discussion, clear communication, careful documentation, and a cooperative attitude will prevent or eliminate most sources of misunderstanding. Careful screening of clients, written agreements, and the implementation of a retainer policy will prevent most bad debts. An attitude of mutual trust, scrupulous observance of professional ethics, and competent performance of your professional engineering services will virtually guarantee client satisfaction.

ETHICS IN CONSULTING ENGINEERING

As a consulting engineer you will be involved in many different types of professional relationships. These include relationships with clients, competitors, employees, contractors and suppliers, and the engineering profession as a whole, as well as with the public. All of these relationships deserve your honesty, trustworthiness, and professionalism. However, each type of relationship has its particular requirements for ethical conduct.

CLIENTS Your first ethical obligation toward your clients is to keep your promises and to live up to the terms of your agreement. You are obliged by your contract to provide a specified level of service at a specified time for a specified cost. Any deviation from the contract must be agreed to by the client.

Depending on the fee structure you are using, various situations may arise in which you must make a decision based on your ethical principles. For example, if you take longer to finish a project than you originally estimated, are you justified in charging the client for the extra time? If the client is not responsible for the delay (for example, by not providing necessary information promptly), then you must either obtain the client's agreement to the extra charges, or be prepared to absorb the difference as overhead cost.

In the reverse situation, where the project takes less time than originally estimated, can you charge the full fee? If the fee is based on time or direct cost reimbursement, you should not. If the fee is a lump sum or other fixed-price fee, however, you may charge the full amount without hesitation.

Your fee should be the only source of income from a given project. You should refuse any trade commissions, discounts, or other types of inducements or compensations that may be offered by the client, suppliers, or other parties.

You may be required to represent your client in your capacity as consultant in communication with other design professionals or with contractors and suppliers. In these situations, you will be entrusted with the client's image and interests, which you must protect even more carefully than your own. At the same time, you must preserve fair and just relations with contractors and suppliers on behalf of the client.

In the course of doing business with a client, it is very likely that you will acquire information about the client that should be considered confidential. You should treat such information as the private property of the client, and not disclose it to any third party. Neither should you disclose to the client confidential information about former employers, former clients, or others. Not only would this violate your obligations to these other parties, it would also demonstrate to the client that you cannot be trusted with confidential information.

If there is any possibility that there will be a conflict of interest as a result of taking on a particular assignment, you must inform the client of the nature of the conflict. For example, if you have any affiliation with a competitor of the client's, or if you have a financial interest in a company that supplies equipment that may be used in the project, the client should be informed of such connections.

If your facilities, staff, or experience are insufficient to assure adequate service on some aspect of a project, it is your obligation to obtain external assistance that is acceptable to the client. This may take the form of consulting, subcontracting, or single-project association, depending on the nature of the problem.

If the client wishes to make changes or decisions that, in your professional judgment, will compromise the success of the project, you are responsible for making the consequences clear to the client. Also, if you have good reason to believe that a project cannot succeed, you must honestly inform the client of your judgment and the reasons for it.

When the client is in a foreign country, you have extra obligations. For instance, you should not undertake an international assignment unless you or someone on your staff has adequate knowledge of the language and society of the country in which the project is to be carried out. Also, while in the foreign country, you (and your staff and associates) should, as guests in that country, maintain the highest standards of conduct. Do not assume that, because people speak a different language, they are any less intelligent than you.

COMPETITORS Competition among consulting engineers should be on the basis of quality of service. Fee-cutting is not only unfair to other engineers, but is unlikely to generate much business for you. Under no circumstances should you say or do anything to harm the reputation of another consulting engineer. If you feel that there is a question about a certain individual's competence or ethical practices, you may wish to discuss the problem with the individual involved, or you may report it to the disciplinary committee of that person's professional society or to your state's registration board. You should never discuss such matters with clients, potential clients, or other third parties.

EMPLOYEES The primary ethical issue with regard to employees is that of fairness. Salaries and other benefits should be adequate and commensurate with the services provided by the employees. Every employee, regardless of rank, should receive fair and impartial treatment and respect for his or her personal dignity.

Personal information provided by employees should be maintained in confidence. Confidentiality should be maintained even if an employee resigns or if his employment must be terminated.

CONTRACTORS AND SUPPLIERS If you are responsible for the purchase of materials or the contracting of services related to a client's project, it will be up to you to obtain these goods and services in a manner that is fair to the bidders. You should never specify equipment or material in exchange for any consideration (including engineering or design services). Nor should you accept compensation of any kind from suppliers or contractors that you engage.

In carrying out the work of the project, you have a number of obligations to the contractors. The plans or specifications you supply must be sufficiently clear and specific for a contractor to be able to follow them. If interpretation is needed, you should supply it promptly and willingly, in a spirit of cooperation. Likewise, you must give the contractor any other information that may be necessary for correct performance of his work. You may not require the contractor to perform services or supply materials that are not specified in the contract, but you are responsible for ensuring that the contractor meets all the requirements of the contract, plans, and specifications.

THE ENGINEERING PROFESSION

As a member of the consulting engineering profession, you have a dual obligation to the profession. You must avoid damage to the reputation of the profession by maintaining professional standards with regard to both engineering competence and ethical conduct, and you should actively contribute to the improvement of the profession and its standing. The maintenance of professional standards involves a strict adherence to competent performance of every aspect of engineering service, business ethics, and the preservation of your professional integrity.

You may make positive contributions to the profession in many ways. For example, you may write technical papers or articles or otherwise engage in exchange of technical information. You may also work toward the improvement of engineering education and try to help engineering students and young engineers. You should recognize other qualified engineers as colleagues in the profession, and extend to them the appropriate considerations and courtesies. You should always be aware that your own reputation is inextricably linked with the reputation of the profession as a whole; you owe it both to yourself and to the profession to maintain your reputation and that of the engineering profession at the highest possible level.

THE PUBLIC

The first obligation of any engineer, whether in consulting or some other employment situation, is to protect the public interest. No matter what your specialty, the work you do is certain to affect the lives of many people. It is your first responsibility to be sure that your work protects the health, welfare, and safety of the public, and that it contributes positively to comfort, efficiency, and other values associated with engineering work.

BIBLIOGRAPHY

American Consulting Engineers Council. *The Federal Client . . . The How, What and Who of Government Architect-Engineer Contracting.* 1983.

American Consulting Engineers Council. *The Federal Client . . . The How, What and Who of Government Architect-Engineer Contracting, Addendum.* 1984.

ACEC Publications List, July 1984.

ACEC Publications List, January 1985.

Adams, Paul. *The Complete Legal Guide for Your Small Business.* New York: Wiley, 1982.

Baird, Michael L. *Engineering Your Start-Up.* Belmont: Professional Publications, Inc., 1992.

CEAC & SEAC, *Guide for Consulting Structural Engineering Services.* June, 1979.

Cohen, Stanley (ed.), *Consulting Engineering Practice Manual.* New York: McGraw-Hill, 1982.

Consulting Engineering: An Overview. *ACEC Guidelines to Practice*, Volume I, No. 1. American Consulting Engineers Council, 1984.

Consulting Engineers and Construction Phase Services. *ACEC Guidelines to Practice*, Volume V, No. 1. American Consulting Engineers Council, 1982.

Cook, Jane T. "Women: The Best Entrepreneurs." *Canadian Business*, June 1982.

Cox, James R. & Karger, Delmar W. "How to Set up a Consulting Practice." *Machine Design*, October 31, 1974, pp. 50-54.

Dunham, Clarence W. & Young, Robert D. *Contracts, Specifications, and Law for Engineers.* New York: McGraw-Hill, 1958.

Engineers' Council for Professional Development, *Suggested Guidelines for Use with the Fundamental Canons of Ethics.*

Engineers' Council for Professional Development, *Code of Ethics of Engineers.*

Engineers' Joint Contract Documents Committee. *Suggested Listing of Duties, Responsibilities and Limitations of Authority of Resident Project Representative.* NSPE No. 1910-1-A, 1983.

Engineers' Joint Contract Documents Committee. *Standard Form of Agreement Between Owner and Engineer for Professional Services.* EJCDC No. 1910-1, 1984.

Engineers' Joint Contract Documents Committee. *Standard General Conditions of the Construction Contract.* No. 1910-8, 1983 Edition.

Engineers' Joint Contract Documents Committee. *Standard Form of Letter Agreement Between Owner and Engineer for Professional Services (With General Provisions Attached).* No. 1910-2, 1979 Edition.

Firmage, D. Allan. *Modern Engineering Practice: Ethical, Professional, and Legal Aspects.* New York: Garland STPM Press, 1980.

Hinett, Virginia D. "Secretarial Specifications." *Consulting Engineer,* October 1984, ppp. 96-97.

Holtz, Herman. *How to Succeed as an Independent Consultant.* New York: Wiley, 1983.

"How Do *You* Define 'Responsible Charge'?" Unsigned article, *Consulting Engineer,* October 1984, pp. 94-95.

"How to Work with Consulting Engineers." Reprinted from *Consulting Engineer* by Technical Publishing, undated.

"The Industrial Market for Consulting Engineers." *ACEC Guidelines to Practice,* Volume VI, No. 3. American Consulting Engineers Council, 1983.

Kemper, John Dustin. *The Engineer and His Profession.* New York: Holt, Rinehart & Winston, 1967.

Kishel, Gregory F. & Patricia Gunter Kishel. *How to Start, Run, and Stay in Business.* New York: Wiley, 1981.

Martin, C. Bernard. "Are Ethics Only for Dealing with Clients?" *Consulting Engineer*, March 1978, pp. 30-34.

"Moonlighting Termed a Definite Employer Danger." Unsigned news item, *Consulting Engineer*, July 1978, p. 14.

"National Society of Professional Engineers: Professional Engineers in Private Practice." *So You Want to Open a Consulting Office.* NSPE Publication #1909.

National Society of Professional Engineers, *Code of Ethics for Engineers.*

Ostrower, Donald A. "Practice in Another State: Problems and Pitfalls—Part 1." *Consulting Engineer*, November 1977, pp. 42-48.

Ostrower, Donald A. "Practice in Another State: Problems and Pitfalls—Part 2." *Consulting Engineer*, December 1977, pp. 14-20.

Ostrower, Donald A. "The NSPE Decision: No Surprise, No Calamity." *Consulting Engineer*, July 1978, pp. 24-28.

Ostrower, Donald A. "Engineering Persons vs. Entities Creates Legal Maze." *Consulting Engineer*, February, 1979, pp. 58-60.

Park, William R. "Starting a New Business?" *Consulting Engineer*, August 1978, pp. 16-20.

Park, William R. "Characteristics of the Independent Businessman." *Consulting Engineer*, September 1978, pp.18-20.

Professional Engineering Institute. "Glossary of P.E. Terms."

Professional Publications, Inc. *How to Become a Professional Engineer.* 1985.

Rowland, Robert D. "Contract Documents: Standard General Conditions, Other Documents, Revised, Added." *Consulting Engineer*, April 1984, pp.110-114.

Silver, A. David. *Up Front Financing: The Entrepreneur's Guide.* New York: Wiley, 1982.

Stanley, C. Maxwell. *The Consulting Engineer.* 2nd Edition. New York: Wiley, 1982.

Sunar, D.G. *How to Become a Professional Engineer.* Belmont: Professional Publications, Inc., 1985.

Tracy, John A. *How to Read a Financial Report: Wringing Cash Flow and Other Vital signs Out of the Numbers.* 2nd Edition. New York: Wiley, 1983.

Weil, Andrew Warren. "Marketing Professional Services." *Consulting Engineer*, November 1984, pp. 90-92.

"The World of Consulting Engineering." *Consulting Engineer*, November 1978, pp. 68-96.
     Tietz, S.B., England: Engineers to the World. Andersen, M. Folmer, Denmark: A Self-Regulating Profession. Walter, H., West Germany: Engineering in a Structured Society. Moulin, Albert, France: Competition from the State. Kawano, Yasuo, Japan: A Profile of the Profession. Edmunds, Jane, International: Design Review and Forecast.

APPENDIX A

A PARTIAL LIST
OF
NATIONAL ENGINEERING SOCIETIES
AND
OTHER ENGINEERING-RELATED ORGANIZATIONS

AACE: American Association of Cost Engineers. P.O. Box 1557, Morgantown, West Virginia 26507-1557. Telephone: (304) 296-8444. Fax: (304) 291-5728.

AAEE: American Academy of Environmental Engineers. 130 Holiday Court, Suite 100, Annapolis, Maryland 21401. Telephone: (410) 266-3311. Fax: (410) 266-7653.

AAES: American Association of Engineering Societies. A private, nonprofit organization to which all of the major engineering societies belong. AAES advises, represents, and fosters communication among the societies. 1111 19th Street, N.W., Suite 608, Washington, D.C. 20036. Telephone: (202) 296-2237. Fax: (202) 296-1151.

ABET: Accreditation Board for Engineering and Technology. A nonprofit, independent organization which is the primary accrediting organization of engineering degree programs. 345 East 47th Street, New York, New York 10017-2397. Telephone: (212) 705-7685. Fax: (212) 838-8062.

ACEC: American Consulting Engineers Council. 1015 15th Street, N.W., Washington, D.C. 20005. Telephone: (202) 347-7474. Fax: (202) 898-0078.

AIAA: American Institute of Aeronautics and Astronautics. 370 L'Enfant Promenade, S.W., Washington, D.C. 20024. Telephone: (202) 646-7400. Fax: (202) 646-7508.

AIChE: American Institute of Chemical Engineers. 345 East 47th Street, 12th Floor, New York, New York 10017. Telephone: (212) 705-7338. Fax: (212) 752-3297.

AIMMPE: American Institute of Mining, Metallurgical, and Petroleum Engineers. 345 East 47th Street, 14th Floor, New York, New York 10017. Telephone: (212) 705-7695. Fax: (212) 371-9622.

AIPE: American Institute of Plant Engineers. 8180 Corporate Park Drive, Suite 305. Cincinnati, Ohio 45242. Telephone: (513) 489-2473. Fax: (513) 247-7422.

ANS: American Nuclear Society. 555 N. Kensington, LaGrange Park, Illinois 60525. Telephone: (708) 352-6611. Fax: (708) 352-0499.

ASAE: American Society of Agricultural Engineers. 2950 Niles Road, St. Joseph, Michigan 49085-9659. Telephone: (616) 429-0300. Fax: (616) 429-3852.

ASCE: American Society of Civil Engineers. 345 East 47th Street, New York, New York 10017. Telephone: (212) 705-7496. Fax: (212) 980-4681.

ASCET: American Society of Certified Engineering Technicians. P.O. Box 371474, El Paso, Texas 79937. Telephone: (915) 591-5115.

ASEE: American Society for Engineering Education. 1818 N Street, N.W., Suite 600, Washington, D.C. 20036. Telephone: (202) 331-3500. Fax: (202) 265-8504.

ASHRAE: American Society of Heating, Refrigerating, and Air Conditioning Engineers. 1791 Tullie Circle, N.E., Atlanta, Georgia 30329. Telephone: (404) 636-8400. Fax: (404) 321-5478.

ASME: American Society of Mechanical Engineers. 345 East 47th Street, New York, New York 10017. Telephone: (212) 705-7722. Fax: (212) 705-7674.

ASNT: American Society of Nondestructive Testing. 1711 Arlington Lane, Columbus, Ohio 43228-0518. Telephone: (614) 274-6003. Fax: (614) 274-6899.

ASPE: American Society of Plumbing Engineers. 3617 Thousand Oaks Boulevard, #210, Westlake, California 91362. Telephone: (805) 495-7120. Fax: (805) 495-4861.

ASQC: American Society for Quality Control. Physical: 611 East Wisconsin Avenue, Milwaukee, Wisconsin 53202. Mailing: P.O. Box 3005, Milwaukee, Wiconsin 53201-3005. Telephone: (414) 272-8575. Fax: (414) 272-1734.

ASSE: American Society of Safety Engineers. 1800 East Oakton Street, Des Plaines, Illinois 60018. Telephone: (708) 692-4121. Fax: (708) 296-3769.

AWMA: Air and Waste Management Association. P.O. Box 2861, Pittsburgh, Pennsylvania 15230. Telephone: (412) 232-3444. Fax: (412) 232-3450.

AWS: American Welding Society. 550 LeJeune Road, N.W., Miami, Florida 33126. Telephone: (305) 443-9353. Fax: (305) 443-7559.

IEEE: Institute of Electrical and Electronic Engineers. 345 East 47th Street, 15th Floor, New York, New York 10017-2394. Telephone: (212) 705-7900. Fax: (908) 981-9667.

IIE: Institute of Industrial Engineers. 25 Technology Park/Atlanta, Norcross, Georgia 30092. Telephone: (404) 449-0461. Fax: (404) 263-8532.

IoPP: Institute of Packaging Professionals. 481 Carlisle Drive, Herndon, Virginia 22070-4823. Telephone: (703) 318-8970.

NCEES: National Council of Examiners for Engineering and Surveying. The organization that writes and distributes the uniform FE and PE examinations. P.O. Box 1686, Clemson, South Carolina 29633. Telephone: (803) 654-6824. Fax: (803) 654-6033.

NICET: National Institute for Certification in Engineering Technologies. An independent, nonprofit certifying organization sponsored by NSPE that examines engineering technicians and awards certificates of competence. 1420 King Street, Alexandria, Virginia 22314. Telephone: (703) 684-2835. Fax: (703) 836-4875.

NSPE: National Society of Professional Engineers. A national professional society encompassing all state PE societies. NSPE represents the social, economic, and political interests of all engineers in the United States. 1420 King Street, Alexandria, Virginia 22314. Telephone: (703) 684-2800. Fax: (703) 836-4875.

Professional Engineering Institute: A nonprofit educational organization that provides assistance, study materials, and state-approved review courses to engineers seeking registration. 850 O'Neill Avenue, Belmont, California 94002. Telephone: (415) 593-9731. Fax: (415) 593-9733.

SFPE: Society of Fire Protection Engineers. 1 Liberty Square, Boston, Massachusetts 02109-4825. Telephone: (617) 482-0686. Fax: (617) 482-8184.

SME: Society of Manufacturing Engineers. P.O. Box 930, One SME Drive, Dearborn, Michigan 48121. Telephone: (313) 271-1500. Fax: (313) 271-2861.

APPENDIX B

ADDRESSES OF STATE
PROFESSIONAL ENGINEERING SOCIETIES

ALABAMA

Alabama Society of Professional Engineers
1150 10th Avenue South, Suite 255
Birmingham, Alabama 35294
Telephone: (205) 934-8470

ALASKA

Alaska Society of Professional Engineers
2064 Belair Drive
Anchorage, Alaska 99517
Telephone: (907) 277-6855

ARIZONA

Arizona Society of Professional Engineers
24 West Camelback Road, Suite M
Phoenix, Arizona 85013
Telephone: (602) 264-4871

ARKANSAS

Arkansas Society of Professional Engineers
920 West Second Street, Suite 102
Little Rock, Arkansas 72201
Telephone: (501) 376-4128

CALIFORNIA

California Society of Professional Engineers
1005 12th Street, Suite J
Sacramento, California 95814
Telephone: (916) 422-7788

COLORADO

Professional Engineers of Colorado
2755 South Locust Street, Suite 214
Denver, Colorado 80222
Telephone: (303) 756-8840

CONNECTICUT
    Connecticut Society of Professional Engineers
    2600 Dixwell Avenue
    Hamden, Connecticut 06514
    Telephone: (203) 281-4322

DELAWARE
    Delaware Association of Professional Engineers
    2005 Concord Pike
    Wilmington, Delaware 19803
    Telephone: (302) 577-6500

DISTRICT OF COLUMBIA
    District of Columbia Society of Professional Engineers
    614 H Street, N.W., Room 923
    Washington, DC 20001
    Telephone: (202) 727-7454

FLORIDA
    Florida Engineering Society
    125 South Gadsden Street
    P.O. Box 750
    Tallahassee, Florida 32302
    Telephone: (904) 224-7121

GEORGIA
    Georgia Society of Professional Engineers
    One Park Place, Suite 226
    1900 Emery Street, N.W.
    Atlanta, Georgia 30318
    Telephone: (404) 355-0177

GUAM
    Guam Society of Professional Engineers
    P.O. Box EK
    Agana, Guam 96910
    Telephone: 011-(671) 339-4100

HAWAII
    Hawaii Society of Professional Engineers
    Walker Industries, Ltd.
    P.O. Box 1568
    Kahului, Hawaii 96732
    Telephone: (808) 877-3430

IDAHO

Idaho Society of Professional Engineers
600 South Orchard, Suite B
Boise, Idaho 83705-1242
Telephone: (208) 345-1730

ILLINOIS

Illinois Society of Professional Engineers
1304 South Lowell Avenue
Springfield, Illinois 62704
Telephone: (217) 544-7424

INDIANA

Indiana Society of Professional Engineers
1810 Broad Ripple Avenue
Indianapolis, Indiana 46220
Telephone: (317) 255-2267

IOWA

Iowa Engineering Society
1000 Walnut Street, Suite 102
West Des Moines, Iowa 50309
Telephone: (515) 284-7055

KANSAS

Kansas Engineering Society
P.O. Box 477
627 S.W. Topeka Avenue
Topeka, Kansas 66601
Telephone: (913) 233-1867

KENTUCKY

Kentucky Society of Professional Engineers
160 Democrat Drive
Frankfort, Kentucky 40601
Telephone: (502) 695-5680

LOUISIANA

Louisiana Society of Professional Engineers
P.O. Box 2683
1213 Nicholson Drive
Baton Rouge, Louisiana 70821
Telephone: (504) 344-4318

MAINE
>   Maine Society of Professional Engineers
>   RR #2, Box 5760
>   Oxford, Maine 04270
>   Telephone: (207) 998-2730

MARYLAND
>   Maryland Society of Professional Engineers
>   720 Light Street
>   Baltimore, Maryland 21230
>   Telephone: (301) 752-3318

MASSACHUSETTS
>   Massachusetts Society of Professional Engineers, Inc.
>   555 Huntington Avenue
>   Roxbury, Massachusetts 02115
>   Telephone: (617) 442-7745

MICHIGAN
>   Michigan Society of Professional Engineers
>   P.O. Box 15276
>   Lansing, Michigan 48901-5276
>   Telephone: (517) 487-9388

MINNESOTA
>   Minnesota Society of Professional Engineers
>   Professional Building
>   555 Park Street, Suite 130
>   St. Paul, Minnesota 55103
>   Telephone: (612) 292-8860

MISSISSIPPI
>   Mississippi Engineering Society
>   5425 Executive Place, Suite D
>   Jackson, Mississippi 39206
>   Telephone: (601) 366-1312

MISSOURI
>   Missouri Society of Professional Engineers
>   330 East High Street, Second Floor
>   Jefferson City, Missouri 65101
>   Telephone: (314) 636-4861

MONTANA
Montana Society of Engineers
1629 Avenue D
P.O. Box 20996
Billings, Montana 59104
Telephone: (406) 259-7300

NEBRASKA
Nebraska Society of Professional Engineers
521 South 14th, Suite 104
Lincoln, Nebraska 68508
Telephone: (402) 476-2572

NEVADA
Nevada Society of Professional Engineers
2835 S. Jones, Suite 5
Las Vegas, Nevada 89102
Telephone: (702) 252-5001

NEW HAMPSHIRE
New Hampshire Society of Professional Engineers
P.O. Box 1343
Concord, New Hampshire 03302
Telephone: (603) 625-8201

NEW JERSEY
New Jersey Society of Professional Engineers
150 West State Street
Trenton, New Jersey 08608
Telephone: (609) 393-0099

NEW MEXICO
New Mexico Society of Professional Engineers
1615 University Boulevard, N.E.
Albuquerque, New Mexico 87102
Telephone: (505) 247-9181

NEW YORK
New York State Society of Professional Engineers
150 State Street, Third Floor
Albany, New York 12207
Telephone: (518) 465-7386

NORTH CAROLINA
Professional Engineers of North Carolina
4000 Wake Forest Road, Suite 108
Raleigh, North Carolina 27609
Telephone: (919) 872-0683

NORTH DAKOTA
North Dakota Society of Professional Engineers
P.O. Box 1031
Grand Forks, North Dakota 58206-1031
Telephone: (701) 777-3782

OHIO
Ohio Society of Professional Engineers
445 King Avenue, Room 103
Columbus, Ohio 43201
Telephone: (614) 424-6640

OKLAHOMA
Oklahoma Society of Professional Engineers
Oklahoma Engineering Center
201 N.E. 27th Street, Room 125
Oklahoma City, Oklahoma 73105
Telephone: (405) 528-1435

OREGON
Professional Engineers of Oregon
1930 N.W. Lovejoy
Portland, Oregon 97209
Telephone: (503) 228-2701

PANAMA CANAL ZONE
Panama Canal Society of Professional Engineers
P.O. Box 6-4455, El Dorado
Panama, Republic of Panama
Telephone: 011-(507) 56-6742

PENNSYLVANIA
Pennsylvania Society of Professional Engineers
4303 Derry Street
Harrisburg, Pennsylvania 17111
Telephone: (717) 561-0590

PUERTO RICO
>   Puerto Rico Society of Professional Engineers
>   Lich Engineering, Inc.
>   GPO 1614
>   San Juan, Puerto Rico 00936
>   Telephone: (809) 722-7272

RHODE ISLAND
>   Rhode Island Society of Professional Engineers
>   9 Newman Avenue
>   East Providence, Rhode Island 02916
>   Telephone: (401) 434-2332

SOUTH CAROLINA
>   South Carolina Society of Professional Engineers
>   P.O. Box 11937
>   1427 Pickens Street
>   Columbia, South Carolina 29211
>   Telephone: (803) 771-4271

SOUTH DAKOTA
>   South Dakota Engineering Society
>   P.O. Box 2004
>   Rapid City, South Dakota 57709
>   Telephone: (605) 394-6674

TENNESSEE
>   Tennessee Society of Professional Engineers
>   530 Church Street, Suite 503
>   Nashville, Tennessee 37219
>   Telephone: (615) 242-2486

TEXAS
>   Texas Society of Professional Engineers
>   P.O. Box 2145
>   3501 Manor Road
>   Austin, Texas 78768
>   Telephone: (512) 472-9286

UTAH
>   Utah Society of Professional Engineers
>   c/o Christensen Environmental Engineering
>   4446 West 1730 South
>   Salt Lake City, Utah 84130
>   Telephone: (801) 535-4144

VERMONT
    Vermont Society of Professional Engineers
    P.O. Box 187
    Montpelier, Vermont 05601
    Telephone: (802) 862-9836

VIRGINIA
    Virginia Society of Professional Engineers
    Heritage Building, Suite 625
    1001 E. Main Street
    Richmond, Virginia 23219-3536
    Telephone: (804) 780-2491

WASHINGTON
    Washington Society of Professional Engineers
    12828 Northrup Way, Suite 300
    Bellevue, Washington 98005
    Telephone: (206) 885-2660

WEST VIRGINIA
    West Virginia Society of Professional Engineers
    179 Summers Street, Room 804
    Charleston, West Virginia 25301-2131
    Telephone: (304) 346-2100

WISCONSIN
    Wisconsin Society of Professional Engineers
    6425 Odana Road
    Madison, Wisconsin 53719
    Telephone: (608) 274-8555

WYOMING
    Wyoming Society of Professional Engineers
    2410 Pioneer Avenue
    Cheyenne, Wyoming 82001
    Telephone: (307) 637-8422